A Volume in
Exploring Complexity

Volume Three
Origins of Self-Organization, Emergence and Cause

Exploring Complexity Series

Volume 1: **Reframing Complexity: Perspectives from the North and South**
Fritjof Capra, Alicia Juarrero, Pedro Sotolongo, & Jacco van Uden (eds.)
ISBN 9780976681465.

Volume 2: **Classic Complexity: From the Abstract to the Concrete**
Kurt A. Richardson & Jeffrey A. Goldstein (eds.)
ISBN 9780979168833.

Volume 3: **Origins of Self-Organization, Emergence and Cause**
Vincent Vesterby
ISBN 9780981703206.

Volume 4: **Emergence, Complexity, and Self-Organization: Precursors and Prototypes**
Alicia Juarrero & Carl A. Rubino (eds.)
ISBN 9780981703213.

Exploring Complexity: Volume Three

Origins of Self-Organization, Emergence and Cause

Written by
Vincent Vesterby

ISCE
Publishing

17947 W Porter Ln
Goodyear, AZ 85338

Cover photo: © 2008 Bors Vesterby. The cover photograph shows the galleries of a carpenter ant colony that were exposed when the log was cut. The circular growth rings and the crosswise cracks determined in large part the pattern of the galleries created by the ants. While the ants followed the set pattern of rings and cracks, they also contributed their own input by cutting cross tunnels here and there. This demonstrates the universal principle that the existence and intrinsic nature of what goes before determines the existence and intrinsic nature of what follows.

Origins of Self-Organization, Emergence and Cause
Exploring Complexity Book Series: Volume 3
Written by: Vincent Vesterby
Library of Congress Control Number: 2008926776

ISBN: 0-9817032-0-8
ISBN13: 978-0-9817032-0-6

Copyright © 2008 ISCE Publishing, 17947 W Porter Ln, Goodyear, AZ 85338, USA

All rights reserved. No part of this publication may be reproduced, stored on a retrieval system, or transmitted, in any form or by any means, electronic, mechanical, photocopying, microfilming, recording or otherwise, without written permission from the publisher.

Printed in the United States of America

CONTENTS

Prologue 1
Preface 3

Chapter 1
Introduction

A Short Description of Emergence 7
Methodology 11
    Epistemology 11
    The Prime Imperative of Analysis 13
    Structural Logic 14
    Development 17
        Extensional Development and Change
            Development Occur by Way of
            Structural Logic 19
    Mental Model 22
    Additional Notes on Method 27
        The Mental Model Is Not a Theory 27
        Reduction 27
        Prediction 30
        Something 30
        Role 30
        Defining 30
        Order of Presentation 31
        History of the Developing
            Understanding of Emergence 32
        Errors and Contrary Opinion about
            Emergence Are Not Discussed 32
Deeper Origins—Foundations, Developments,
    Interrelations, and Origins That Are
    Basic to the Origin of Emergence 32
    Reality 33

Chapter 2
Space

Space Exists—Space Is Place 37
    Spatial Place Is Immaterial 37
    Spatial Place Has Extension 37
    Spatial Place Has Parts 39
    Relation 40
    Combinatorial Enhancement 41
    Sequentiality—Sequential-Difference 42
    Extensional Development—
        Sequential Enhancement 44
    Spatial Place, Existential Organization, and
        the Foundation of Pattern 47
    Existential Context—A Place-to-Be 48
    Existential Requisites, and the
        Intrinsic Self-Identity of Space 49
Spatial Place Exists, and It Continues to Exist 50
    Continuance, Quantity, Parts 50
    Sequentiality, Sequential-Difference,
        Noncoexistent-Sequential-Difference ...51
    Relative Aspects of Noncoexistent-
        Sequential-Difference 52
    New Part, Change, Change Development 53
    Change in an Aspect of Self-Identity 54
    Consequent-Existence and
        Determinate-Reality 56
        Consequent-Existence............................56
        Determinate-Reality................................57
    Existential-Dependency 59
    Uniformity and Unidirectionality of
        Continuing-Existence 60
    Continuing-Existence and Organization 60
    Spatial Continuance-of-Being Has Always
        Occurred and Will Always Occur 61
    Simultaneity 62

    Spatial Continuing-Existence
        Existential-Context 62
The Primal Development from the Existence of
Space to the Continuing-Existence of Space 64
The Initiators 66
    Initiation and Intrinsic Consequence 66
    Determinate Aspect of Initiation 67
    Development of Initiators 67
    Initiators and Sequential-Difference 68
The Continuing-Existence of Space Provides the
    Temporal Aspect of Reality 69
    Time—In Summary 73

Chapter 3
Space and Matter

Using Space to Understand Matter 78
    Factors of the First Stage Required for the
        Existence of Matter 79
    Factors of the Second Stage (Continuing-
        Existence) Required for the
        Existence of Matter 84
The Identity of Matter Is Based on the Existence
    of Substantiality 86
Materially Based Extensional Development and
    Materially Based Change Development 87
Space and Matter Are Coexistent—
    Matter Occupies Space 90
Development of Factors from Their Spatial Form
    to Their Material Form—Nonpathway
    Factor Development 95
    Preliminary Considerations 95
    Nonpathway Factor Development from
    Space to Matter 97

Chapter 4
Space, Matter, and Motion—The Origin of Emergence

The Existence of Motion 103
   The Intrinsic Nature of Motion 103
      Factors of Self-Identity and
         Continuing-Existence of Motion ... 103
      Factors of Motion Itself—The Basic
         Intrinsic Qualities of Motion 105
   Motion Is an Initiator 110
      Initiation Situation—
         Organizational Initiation 112
   Motion Is a Developed Form of Change 119
   The Role of Isomorphism 120
   The Role of Continuing-Existence in
      Developed Change 123
Space and Two Static Units of Matter 125
Motion of a Unit of Matter in the Presence of
   Another Unit That is Not Moving 129
   Factors of This Stage 130
   Discussion of Factors at the Origin of
      Emergence 133
Why Is There Emergence? 138

Chapter 5
The Origin of Cause

The Development That Leads to Cause 142
   A Review of the Development of
      Initiation 142
   The General Pattern of the Development 144
      Factors That Determine This
         General Pattern of Development 145
      Development at Third Level Extensional
         Pattern 146

  The First Stage of Initiation at the
   Third Level of Extensional
   Pattern 146
  The Second Stage of Initiation at the
   Third Level of Extensional
   Pattern 146
  The Third Stage of Initiation at the
   Third Level of Extensional
   Pattern 150
  Additional Points of Interest about
   Transformation Points 152
 Development at Fourth Level Extensional
  Pattern 153
  The First Stage of Initiation at the
   Fourth Level of Extensional
   Pattern 154
  The Second Stage of Initiation at the
   Fourth Level of Extensional
   Pattern 155
  The Third Stage of Initiation at the
   Fourth Level of Extensional
   Pattern—And the Emergence
   of Contact 158
The Development-of-Origin of Cause 164
  The Third Stage of Initiation at the
   Fourth Level of Extensional
   Pattern—And the Emergence
   of Push 164

Chapter 6
The Origin of Self-Organization

The Basics 172
Self-Organization through the Collision
  Situation 181
  Adjacent Relation 182
  Contact Relation 182
  Blocking 184

The Changed Motion of the Moving Unit 189
Push 191
The New Motion of the Blocking Unit 192

Appendix 1
The Differences between the Extension of Spatial Place and the Continuing-Existence of Spatial Place 197

Appendix 2
Summary of Factors That Are Identical in Spatial and Material Continuing-Existence 203

Appendix 3
Two Lists of the Factors That Play Roles with Motion 205

Appendix 4
Summaries of the Development of Some of the Individual Factors that Occur in the Foundational Development of Reality 217

Origins of Self-Organization, Emergence and Cause

This book is about how emergence, self-organization, and cause come into existence. These fundamental processes play roles in the origins of virtually everything, thus the book describes the basics of how everything comes into existence. Development plays many roles in the origins of emergence, self-organization, and cause. With factor development, factors such as emergence originate in simple form in simple situations, and occur in more complex form in more complex situations where additional factors are playing roles. With situation development, the interrelations of two to many factors change through time. Development can be creative, leading to progressive increase in complexity. It is a universal factor that provides a way to develop a universal conceptual model. Understanding development, emergence, self-organization, and cause, provides the opportunity to become a modern generalist. A modern generalist thinks in the mode of these factors, using them as intellectual tools of exploration, analysis, understanding, and description.

The modern generalist mode is like a Rosetta Stone of understanding. It translates the intrinsic deep structure of reality into a form that can be comprehended by a living mind.

In the analysis of the intrinsic nature of reality one must expect any situation to be anywhere from ten to thousands of times more complex than first anticipated.

Preface

Origins of Self-Organization, Emergence and Cause is about how things come into existence. Primarily, it is about how emergence comes into existence. It is by way of emergence that newly occurring patterns of material organization come to be. Since material reality above the level of elementary particles is composed of groups of particles, each with its own pattern of organization, the origin of these groups and their patterns is of considerable interest. The book is also about the origin of cause, which is emergent, and it is about the origin of self-organization, which in its developed stages is a form of emergence. Additionally, through the process of exploring and describing these matters, the book came to be about thinking like a generalist.

After the introduction, the following chapters present the identity of emergence at its foundational, simplest form, which plays a role in all other more complex cases of emergence, giving them their intrinsic nature. These chapters explain why emergence exists at all, why it occurs as a creative factor of reality. They also point out that emergence develops, becoming more complex as additional factors play roles in the process of emergence in more complex situations.

There are two basic meanings for the term emergent, (a) to come into view or awareness, visually or conceptually, or (b) to come into existence. Here the term is always used in the latter sense. The first meaning has anthropocentric aspects, while the other does not. Confounding the two meanings has led to a great deal of confusion, so much so that the topic is not well understood. The understanding of the intrinsic nature of emergence has been distorted by anthropomorphic issues of surprise, novelty, prediction, and reduction. The situation is so confused that John Holland had this to say in his book, *Emergence: From Chaos to Order* (1998: 3): "Indeed, our present understanding of emergence is often little better than the child's invocation of Jack Frost to explain the wondrous colors of autumn," and "It is unlikely that a topic as complicated as emergence will submit meekly to a concise definition, and I have no such definition to offer."

Origins of Self-Organization, Emergence and Cause provides an understanding of the core nature of emergence in its simplest, foundational form, and explains why, at more developed stages and higher levels of the organization of reality, the more complicated forms of emergence require more elaborate descriptions.

In addition, the book provides an introduction to generalist techniques of analysis and understanding. I first began the search for generalist techniques through independent study courses in my junior year in college. I did not find them until going through the actual process of writing this book. Emergence originates very early in the development of the organization of reality, and to understand why it exists as a creative factor, it is necessary to go even deeper to understand the intrinsic natures of the factors that together constitute emergence.

It was by exploring the origins of these more fundamental factors, and then following their interwoven developments to the origin of emergence, that I achieved understanding of several other factors of foundational organization, factors that play universal or near universal roles in the integration of the components of reality through space and time. It was not until I was forced to use them as tools of investigation that I came to see their significance as techniques of generalist understanding. The process of going deep, and then reconstituting reality from there, bit by bit, revealed the manner in which these universal factors establish the connections that make reality a structural and functional unity.

Primary among these unifying factors is development, which gives order to virtually all other factors, and by which all other factors can be understood in context. Understanding development results in an integrated progressive and vertically oriented world view from space and time to societies, consciousness, galaxies, and the universe. This is a world view in which the boundaries of the disciplines of science fade in significance, along with their hierarchically layered, horizontally orientated, and limited viewpoints.

Since it is no longer possible to be a generalist in the traditional sense, because of the vast amount of knowledge available, it is necessary to explain what it now means to be a generalist. Traditional generalists were characterized by the nearly all encompassing breadth of their knowledge. Modern generalists are characterized not by how much they know, but by what they know. It is not quantity of knowledge, but kind of knowledge that makes it possible once more to be a generalist.

Even if an individual achieved doctoral degrees in a series of disciplines from physics to sociology, that person would still not be a generalist. The reason is that the degrees would be for disciplines, for disciplinary viewpoints, the viewpoints of specialists. Because of the disciplinary approach, because of the limited nature of the world views of the disciplines, even that great breadth of knowledge would still lack the integrative features of a generalist world view. For like reasons, a generalist understanding cannot be achieved by a committee of specialists.

It is the integrative features that distinguish the generalist approach. A generalist uses those aspects of reality that connect one part with another to achieve a unified world view. The orientation here is progressive through time and vertical up the hierarchic organization of material reality, rather than the comparatively static horizontal orientation of the specialist disciplines. But such a view is only a consequence of generalist procedure. Generalists follow pathways of connection. They do so by tracking these factors:

1. Structural logic;
2. Coexistent-sequential-difference;
3. Factor development;
4. Sequential enhancement;
5. Determinate consequent-existence;
6. Continuing-existence;
7. Self-organization;
8. Noncoexistent-sequential-difference-change;
9. Existential-dependency;
10. Situation development;
11. Motion;
12. Combinatorial enhancement;
13. Emergence;
14. Cause, and;
15. The ongoing interrelation of diverse situation developments.

Origins of Self-Organization, Emergence and Cause describes the intrinsic nature of emergence, cause, and self-organization in their simplest forms, which can then serve as the basis for understanding more developed, complex, higher level forms. Emergence, cause, and self-organization develop, as do other factors of the origin and organization of material pattern. A major benefit of understanding the origin and development of these three factors is an accurate understanding of what patterns of organization occur at the various levels of the hierarchic organization of material reality, from elementary particles to life, ecosystems, planetary systems, and galaxies.

While the realm of reality this book discusses is fundamental, primarily the foundational stages and levels of development, it can be used as a primer for techniques and conceptual approaches for the analysis and understanding of more developed stages and higher levels.

Everything is understandable.
Achievement of understanding is everything.

Chapter 1
Introduction
A Short Description of Emergence

As a factor of the intrinsic nature of reality, existing independently of human knowledge about it, emergence is, at its origin, the universal determinate process of creative change based on consequent-existence by which newly occurring patterns of material organization come into existence.

Setting aside concepts whose validity is not grounded in reality, and concepts whose validity has not been established, such as postulates, hypotheses, and the speculative aspect of theory, patterns of material organization appear to be fundamentally particulate in nature. All known objects, populations of objects, and systems exist as elementary particles or as groups fundamentally composed of such particles. The difference between simple groups, such as atoms, molecules, and rocks, and complex groups, such as geologically dynamic planets, cells, and ecosystems, are their respective sets of component elementary particles and their respective patterns of interrelational organization.

Emergence is creative change because something comes newly into existence, a different pattern of material organization that was not there before. Change is creative because of the factor of newness. There are usually eleven cases of change that play essential roles in the development-of-origin of the process of emergence. Thus, there are usually eleven roles for newness in the origin of the creative change by which new patterns of material organization come into existence. Each of the eleven cases of change and associated newness plays a distinct role that makes emergence possible. One of these eleven provides the existential context for all forms of change. Three are the primary forms of change that separately or in combination provide the foundational source of all change. One occurs as an intrinsic quality of motion. Some play roles in the relations of a moving unit of matter to space, and others play roles in the relations of a moving unit to another unit. These eleven roles of change and consequent newness interrelationally combined constitute the fundamental process of emergence.

Emergence is a process of consequent-existence, the existence of one thing as a consequence of the existence of something else. Emergence occurs because space, matter, continuance-of-being, and motion occur. The motion of matter through space during continuing-existence has the consequence that spatial relations between units of matter change. When the spatial relations between units of matter change, the patterns of organization of those units change. Patterns of organization come newly into existence—they emerge. This situation, this set of factors and their interrelations, constitutes the consequent-existence origin of emergence, the development-of-origin for emergence.

Emergence develops, becoming more complex as it occurs in relation to an increasing variety of factors in increasingly more complex situations. It is typical of the factors that constitute reality, factors of existence, organization, and change, to occur in more complex forms and to play more complex roles as they occur in more complex situations. For example, the intrinsic nature of consequent-existence develops as it plays more developed roles from its origin with (a) continuance-of-being, then through (b) motion, (c) relative change and the origin of emergence, and (d) causal relations, as the various additional factors involved with motion, relative change, and cause begin to play their roles. The factors of consequent-existence in continuance-of-being are fewer in number and simpler in nature than the factors of consequent-existence that play roles in the physical contact interaction of cause. The factors that play roles in emergence at its origin

Chapter 1: Introduction

with change in spatial relations between units of matter are fewer in number and simpler in nature than the factors that play roles in the somewhat more developed stage of causal emergence, and vastly fewer and simpler than the factors that play roles in the highly developed emergence that occurs with biological evolution. Some factors of existence, organization, and change whose developments play significant roles in the origin and initial development of emergence are: existence, organization, initiators, continuance-of-being, change, sequentiality, newness, determinate-reality, consequent-existence, combinatorial enhancement, motion, cause, and development. Because the factor development itself develops, there are several types of development ranging from the development of fundamental factors of existence, change, and organization up to advanced forms like the developments that are biological growth and learning.

Emergence is determinate, intrinsically determinate. During change the nature of what goes before determines the nature of what follows. Emergence, the change that is the coming into existence of newly occurring patterns of material organization, has four particularly important sources of change, and thus four particularly important sources of its determinate nature. The first source of change that plays a role in the determinate nature of emergence is the change that occurs with spatial continuance-of-being, the change that occurs fundamentally as the continuing-existence of space. The second is the change that is motion, the process of matter passing through the continuum of spatial place. The third is the relational changes of distance, direction, and positional orientation between units of matter that occur as a consequence of motion, and which have in turn the consequence that patterns of material organization come newly into existence. This is the origin of emergence itself. The fourth significant source of change that plays a role in the determinate nature of emergence occurs as a stage in the development of emergence when physical contact interaction results in cause. Emergence is determinate because it is a process of change in which what goes before determines what follows.

Emergence is universal. Emergence is not simply the creation of any particular type of pattern or level of material order, such as the level or stage where cause first plays a role, for there are changes in patterns of organization, the actual emergence of newly occurring patterns, that are simpler, developmentally prior, to the emergence of cause. Nor is emergence simply the creation of material order in general. New patterns occur also with the creation of disorder. Emergence, rather, is the creation of any newly occurring pattern of material organization. It is the emergence of new pattern that is foundationally significant, not the type of pattern that is newly existent.

Emergence, as a universal process, has two basic forms, emergence from combining and emergence from separating. Emergence from combining is the emergence pathway of combinatorial enhancement. For example, a sun-like star emerges from the combining of a large quantity of matter. Emergence from separating is the emergence pathway of energy dissipation and matter dispersal. For example, a planetary nebula eventually forms from the spreading out of matter from a sun-like star when it transforms into a white dwarf. Rearrangement is a combination of the two.

Any change in material organization is an event of creative change because there are always the eleven forms of change and their eleven forms of consequent newness. In particular, there is always continuance-of-being, motion, and relational change. The change that is continuance-of-being is constant and universal, and the motion of matter and the consequent relational change between units of matter are virtually so. The organization of the material universe—the organization of all material reality—is constantly emerging.

Methodology

The term emerge has several meanings, one of which is "to come into existence" (Random House Unabridged Dictionary, 2nd ed. 1993). As used here, the terms emerge, emergence, emergent, and emerging, refer to the process and event of coming into being. They are meant to refer not just to the concepts of coming into being, but to the reality referents of those concepts.

Emergence is not a man-made aspect of the universe. It occurs elsewhere—Andromeda, quasars—places where humans have no influence on the nature and course of events. Emergence is real, it happens. Concepts of the nature of reality, and specifically of the nature of knowledge, understanding, and of emergence, that are anthropocentric or anthropomorphic, such as solipsism, idealism, or Platonism, are by their very nature incapable of providing even a rudimentary understanding of the origin or intrinsic nature of emergence.

The accurate analysis of the intrinsic nature of emergence requires a realist approach. What is the intrinsic nature of emergence, why does it exist, and where in the organization of reality are its foundations or origin?

The realist approach that leads to the understandings presented here has five parts. The first is a realist epistemology, a biological epistemology. The second is the guidance of what can be called the prime imperative of the analysis of the intrinsic nature of reality. The third is the use of structural logic as a tool of exploration. The fourth is understanding development. And the fifth is the construction of a nonsymbolic mental model based on structural logic.

Epistemology

Epistemology is the study of the origin, nature, methods, and extension of knowledge. Epistemology sets the conditions for the process of analysis and for understanding. An epistemology, then, that is used to guide an investigation is a summary statement of the origin and nature of knowledge, and derived from that, various tools of analysis and understanding to guide the mind to accurate knowledge.

The understanding of the nature of reality is derived from experience and careful thinking. Experience and thinking are biological factors. The analysis of the nature of reality, and of emergence as an aspect of reality, must be based on a biological understanding of the nature of experience, knowledge, thinking, and understanding, that is, on a biological epistemology.

A human being is a biologically evolved entity, one subunit of which is a biological-thinking-unit that receives data from its environment by way of sensory systems. This biological-thinking-unit uses this data to construct various images and other forms of relatable experience, at which time the data becomes knowledge. These images and other experiences can be manipulated—compared, stored, retrieved, and altered. Through processes of interrelating and manipulating this knowledge gained by experience, the mind can identify relationships and also build relationships among the various parts of experience, thus producing new knowledge useful for the purpose of understanding. Understanding is a higher order of knowledge than basic sensory experience, or the symbolically encoded information possessed by the mind such as simple facts stated in words or represented mathematically. Understanding is the knowledge of interrelationships between things that exist. Lower order forms of knowledge deal with what, while understanding deals with why. Understanding emergence requires understanding relationships between factors, understanding why they interrelate the way they do.

As the brain, with its associated mind, is a biologically evolved entity, it has a functional role in the ongoing process of evolution. It is the central coordinating unit for the cybernetic control of the body and of the body's interrelations with its environment. The brain and the mind play roles in the reproductive processes of the individual organism. For them to operate in a manner such that reproduction is an ongoing process, they must have sufficiently accurate data concerning the external world, and they must be able to interrelate that data in a sufficiently realistic way. The brain, the mind, and their supportive sensory systems have evolved to realistically fit the roles they play in the processes of life and evolution.

For the mind to play its ecological and evolutionary roles in a manner that keeps the process of life ongoing, it must have accurate data about the world in which it exists. Truth is the accurate correspondence between the data, images, ideas, and understandings of the mind and the things they depict, represent, or refer to—their reality referents. Truth is a product of the evolutionary process, and it has its origin in this necessity that there be a strong correspondence between the data, images, ideas, and understandings with which the mind works and the external world in which the mind and its associated body function.

The next few chapters are not about the concept of emergence, they are about the reality referent of that concept. This endeavor is one of exploration, of discovery. It is truth that is sought, an accurate understanding of the intrinsic nature of emergence. It is to that end that the other four parts of the realist approach are directed.

The Prime Imperative of Analysis

The prime imperative of analysis is to look to the subject of investigation itself. Its purpose is to let reality tell its own story, to let the intrinsic nature of reality dictate the nature of the developing understanding of reality. It is a procedure of uninhibited exploration. There should be nothing to prove. The prime imperative directs the attention beyond hopes, preconceptions and expectations, prejudices, fear of what might be found, undisciplined emotions, received opinion (tradition, authority, and consensus), vested interests (time, effort, accomplished work, and self-esteem), cherished beliefs, symbolic representations, etc. To follow the prime imperative is to look to the whole situation, to everything that plays a role in the situation, including its history.

Structural Logic

Structural logic is something that is discovered through the application of the prime imperative. The first step in that application is to just look at the subject of investigation. Let the eyes, the mind, and the other senses roam all over it, sensing what is there. What patterns of existence are there? What interrelations are there between the components of the patterns?

It will be seen that each component of a pattern has its intrinsic qualities, such as what it is made of, its size, and its shape. It can be seen further that these qualities of a component play roles in the manner in which the component interrelates with other components in the pattern. Repeated observation shows these interrelational roles of the intrinsic qualities of the components to be consistent, and it shows why they are so. Sufficiently careful observation reveals that, in association with the manner in which the components come together, the qualities of components dictate the manner in which the components interrelate. Any particular interrelation will not occur, cannot occur, in the absence of the required qualities of the components. And any particular interrelation will occur, under appropriate circumstances, if the required qualities are present in the components.

If there is a smooth board lying flat on a horizontal table, and there is a sphere resting on the board, in the context of gravity and the horizontal position of the board, the sphere will remain at rest. If the end of the board is raised, the sphere will roll down the slope. The end of the board will not have to be raised very far before the sphere will begin to move. In the same situation, if the sphere is replaced by a cube of equal quantity of matter, the cube will not begin to move when the board is tilted to the same degree at which the sphere started rolling. The end of the board will have to be raised much higher, resulting in a much steeper slope, before the cube will begin to move. When the cube does move, it will not roll, but instead will slide down the board.

The qualities of the sphere and the cube, their shapes, dictate when motion will begin and the form that motion takes. If the sphere and the cube are replaced by a round rod and a square rod, the same set of relations will occur. If a pentagonal rod is placed on the board, and the board tilted, the rod will slide much like the cube and square rod. However, if a ten-sided rod is on the board when it tilts, this rod may begin its motion with a slide, but will begin to roll with a greater tilt of the board. The cross-sectional shape of the ten-sided rod is somewhere between that of the round rod and the square one, as is its relation to the degree of tilt.

There is a natural logic, structural logic, to the relation between the qualities of components and the interrelations between components—If this, then that. If round, under the influence of gravity, and on an inclined surface, then roll. If square, under the influence of gravity, and on a sufficiently inclined surface, then slide. The absence of the required qualities makes a particular interrelationship impossible. The presence of the required qualities makes an interrelationship possible. The basics of structural logic are exceedingly simple. Nevertheless, reality exists and operates by way of structural logic.

As an example of the significance and utility of structural logic, consider math. The universal occurrence of quantitative relations, the universal utility of math, and its universal validity, has prompted some to view math as possessing an unreasonable significance in the greater scheme of things. But the significance and utility of math are entirely reasonable, entirely understandable. The reason is that quantitative relations are a form of structural logic. Math derives its nature, its validity, and its utility from its foundation in structural logic.

The role structural logic plays in math constitutes a vanishingly small part of its role in the quantitative relations that exist in the universe. Compare the volume of space actually occupied by math in the minds of sentient-cognitive beings and in their artifacts such as computers and books to the volume of the universe. The role of structural logic in quantitative relations in general constitutes only part of its role in the intrinsic nature and organization of reality. Everything that exists has its intrinsic qualities, quantitative and qualitative, and the interrelations between everything that exists are dictated by those qualities and the manner in which they occur together. Structural logic, then, provides something of a universal roadmap to the structure and function of reality, and can be used as a tool of exploration and prediction.

The mental model, described a little further along, is derived from three features of reality found by applying the prime imperative of analysis to the intrinsic structural logic of reality—factors, development, and interrelation. A factor is that which plays a role in the existence and nature of a situation. Factors are diverse in character. They can be immaterial or material. They can be something that exists, like space or matter, or a material pattern that has continuing-existence in space. They can be something that exists in the manner of qualities, aspects, or features of space, matter, or material pattern. They can be relations between things or other factors, such as a distance relation, or a role something plays in a situation. And they can be change, change development, or qualities thereof, such as noncoexistent sequential relation, existential-dependency, or existential-pathway-development relation. A factor is something that exists, anything that exists, that plays a role in the existence and nature of reality.

In the simple examples of structural logic, the sphere, the cube, the board, the space in which they exist, and gravity are factors of the situation. The roundness of the sphere, the squareness of the cube, the horizontal position of the board and its flatness, the relative positions of the sphere or cube and the board, the tilt of the board, and the degree of tilt are factors of the situation. The motion of the sphere and the cube, the types of motion, when the motion starts, the substantiality of the objects, and their coherence such that they maintain their shapes are factors. Every thing, feature, or aspect of a situation is a factor of that situation.

Development

Factors and situations develop. Development itself develops, resulting in a variety of kinds of development. Although it is not one of the foundational forms of development, combinatorial hierarchic development is one of the easiest types to explain. Consider a cubic crystal. It is composed of atoms, which in turn are composed of elementary particles. The crystal has a combinatorial hierarchic structural organization, its atoms being combinations of elementary particles, and the crystal being a combination of the atoms into the cubic lattice. Eight such cubes can be stacked together to form a larger cube. It is hierarchically constructed from the combination of the smaller cubes.

Two, more fundamental, kinds of development play roles in this case of combinatorial hierarchic development. First, there is situation development in which a group of units, the elementary particles, progressively combine and recombine to produce, in this case, sequentially more complex organization. As an example of a highly developed case of situation development there is biological development from a fertilized egg to a fully grown adult organism. Second, there is factor development in which a factor, cubic shape, occurs first at one stage of a developing situation and then occurs again at a later stage, with the cubic form at the later stage being more complex than the prior form in that it has in its structure an additional hierarchic level. A case of highly developed factor development is the sequence of progressively more complex stages, generations, of the factor, organism, from the single celled stage like bacteria to stages like elephants and humans. Note that the development from the origin of life to any individual living person is an unbroken, several billion year old, situation development. The living process of any individual being alive on this planet today is billions of years old. As the current stage of the factor development, you are not very old—as the current part of the situation development, you are ancient.

Development originates in exceedingly simple fashion as an aspect of the existence of space, and develops into forms as exceedingly complex as the biological evolutionary process or the development of a culture. There are two basic sequences of development, extensional development and change development. Extensional development involves factors of the type studied by geometry, while change development involves factors of the type that play roles in events and processes. Even though these two developmental pathways each originate in a distinct manner wherein the nature of the mode-of-being of the one does not play a role in the nature of the mode-of-being of the other, in the foundational development of reality they soon begin to interrelate.

The interrelational development of reality is multipart, being first the interrelation of factors into a situation, into a whole of some form, and second the interrelation of situations into greater situations, into greater wholes of more complex form, third the interrelation of the developments of factors, and fourth the interrelation of developing situations. The interrelational development of reality is ongoing, constantly creating the current state of reality. The pattern of reality is the product of the manner in which all the various factors interrelate as they develop together. It is possible to develop an understanding of the nature of the pattern of reality by exceedingly careful tracking of the interrelating sequences of development of the various factors and situations. By following the ongoing interrelations of multipart development, it is possible to achieve an understanding of why the current state of reality is what it is.

The sequential order of development can be simplified into a list of the factors and their stages of development. Because such a list provides a summary view of the developmental organization of reality, it is somewhat like a map of that development, and is here called a list map. The Appendix has two lists of the factors that play roles with motion. List mapping is an important tool for keeping track of progressing understanding during the exploration of the diversity and complexity of the developmental aspect of reality.

Extensional Development and Change Development Occur by Way of Structural Logic

The manner in which factors interrelate occurs by way of the structural logic that results from the intrinsic nature of the factors. Development of factors and of situations also occurs by way of structural logic, as do the ongoing interrelations of developing situations. The organization of reality can be followed by way of structural logic from its simplest foundations to its greatest complexities, from its history to its present state, and on to its future. Because emergence is a developmental process whereby something new comes into existence, and because emergence operates by way of structural logic, the understanding of emergence can be used as a tool of exploration and prediction.

The development of extensional relations and the development of change relations occur sequentially. Consider a path across a meadow. From any point on that path, the pathway extends away sequentially in either direction. From a point on the path, the pathway has, sequentially, more and more distance from that point. This is a particularly simple case of extensional development. The use of the term path in this case of extensional development is clear and unambiguous, being in accord with the basic definition of the word.

A development of the meaning of this word occurs in situations where an object moves through space, through some place, such as the meadow. When a bear crosses that meadow following the path, the bear follows the route that is there as the path. When the bear swerves off the path to drink from a stream, the path of the bear follows a different route across part of the meadow. This is a different use of the term path. In the previous case the route was differentiated physically by the existing path. In this case the path is just the route the bear takes through the grass from the established path over to the stream.

In both cases, the established path and the route of the bear through the grass, there is an aspect of sequentiality through space, an aspect of linear extension. One case is materially differentiated, by the compacted soil, but the other case is not so differentiated. The route or path of the bear through the grass is a sequential event through space. It is the pathway of the event that is the bear's motion through the meadow. In this case it is both a sequential path through space and a sequential pathway of change.

The meaning of the term pathway, as a sequential pathway of change, has been further developed to refer to the pathway of a person's life. While the path of a person's motion through space, across meadows and along the streets of cities, is included within this use of the term, the emphasis of meaning here is more the sequence of the events of a person's life. Biochemistry uses the term pathway in this sense of sequential pathway of change. For example, "...a sequence of reactions, usually controlled and catalyzed by enzymes, by which one organic substance is converted to another" (Random House Unabridged Dictionary, 2nd edition, 1993).

In both extensional development and change development, there is a quality of sequentially connected development. There is a role for continuity in the extension of a path across a meadow, in the route of a bear through the grass, in the living of a life, and in the series of steps of an ongoing sequence of chemical reactions. What follows is developmentally related to or even existentially-dependent on what is developmentally prior.

This aspect of existentially connected development is here called existential-pathway-development. When following the development of the organization of reality from its foundations to the various types and levels of complexity, from its history, through to prediction of aspects of the future, the sequences of developmental relations that are followed are existential-pathway-developments. And existential-pathway-developments occur by way of structural logic relations.

The foundation of structural logic is that it is the intrinsic nature of the qualities of factors that make it possible for there to be various kinds of relations between factors. If a factor has the required quality, then a particular relation can occur. If a factor does not have the required quality, then a particular relation cannot occur. The existence of something has the consequence that the existence of something else is possible. The nonexistence of something means that its consequence is not possible.

Reality operates by way of possibilities. It does not operate, it cannot operate, by way of impossibilities. It does not operate by way of the absence of something. Something that is not there cannot play a role in a situation. For example, consider a plant and the absence of a critical factor such as light. In a situation in which all other required factors, such as moisture and warmth, are present, a bean seed will germinate and put up its cotyledons, and perhaps the first few leaves. In the absence of light, however, the plant will stop growing and die. But the plant does not die because there is no light. Nonexistence is not a factor. That which does not exist cannot play a role. The plant dies because it has exhausted the possibilities for its growth in that particular situation. The factors in the situation that have growth as a consequence of their togetherness have all interrelated as far as is possible. If light becomes available to the plant as an added factor, then the combination of the previous group of factors that have growth as their consequence and the light will have the consequence that the plant will continue to grow. The absence of light does not have the consequence that the plant dies, but the presence of light as an additional factor in the situation does have the consequence that the plant continues to grow.

Something cannot come from nothing. Change cannot occur from nothing. Something cannot come into existence from nothing. Something cannot emerge from nothing. By its very nature nothingness lacks the wherewithal for something to come from it.

Absence has no consequences, presence does. Nonexistence has no consequence, existence does.

Therefore, when using structural logic as a tool of exploration, if there exists a consequence or an effect, then there was or still is a prior or simultaneous factor that played or is still playing a role in the existence of that consequence or effect. In general, if a factor exists, there were or still are other factors that played or are still playing a role in its existence. This applies to every factor that exists in all of reality. Reality as a whole is a continuously ongoing situation development.

Absence is not a constraint. Absence has no consequence, plays no role in any situation, and is thus not a factor. Constraints are factors that exist. They play roles that involve blocking the consequences of other factors. The material and shape of a cup prevent the tea within from flowing under the influence of gravity. Knowledge of unwanted consequences of a particular act can play a role within the mind of preventing the initiation of that act. The cup and the knowledge both have a material foundation to their existence, and that matter has in each case a specific set of qualities, for instance the shape of the cup and the pattern of organization of the brain's encoded knowledge, that play roles in constraining the flow of the liquid and the flow of the signals in the brain that would initiate that action.

Mental Model

By observing factors, their interrelation into situations, the developments of factors and situations, and the interrelation of ongoing situation developments, there develops in the mind a mental model of these relations. This is a natural learning process, the result of paying attention

Chapter 1: Introduction

to something. But it soon develops into something more. This continuously growing mental model is an overview of all these relations together, as a whole. It shows the overall organization of these factors, it shows the intrinsic order of reality. The structural logic of the relations and the resulting organization of the mental overview becomes a tool for the placement of additional factors and for identifying new relationships.

Because the organization of the mental model is derived from the structural logic of the interrelations between factors, which is in turn derived from the qualities of the factors, it is necessary, for the model to be fully effective as a tool, that the factors be represented there in a manner such that the structural logic is evident. The factors should occur in the model in forms as close to the intrinsic nature of their reality referents as is mentally possible. Thus, roundness should not occur in the model as the word round, but rather as round, for example, O. Carve the letters of the word round out of wood, glue them together so that they spell round, and put them on a sloping board. They will slide before they will roll. The structural logic associated with roundness is not directly evident with the word. Carve a ring or a ball out of wood and place it on the board. What happens is an obvious consequence of the shape of the object. The components of the mental model should occur there in forms that make the structural logic directly obvious. This model, therefore, cannot be composed of symbols, mathematical formulas, or language.

Simply put, since it is a model of reality, picture reality. This is a natural thing which everyone does. Everyone has some form of mental model of the layout of their home, school, workplace, or community. Additionally, almost everyone imagines situation developments in the form of imaginary discussions or developing social situations. Combining the two results in a mental model of place and ongoing change or situation development.

Just as a person can enter a building, walk around in its hallways and rooms, and open a desk drawer to get a pencil, the mind can float through the mental model of the building and its contents. Walking through a building is a sequential process, a form of situation development. If a person walks through a building, carefully observing along the way, a mental model of the walk is created within the mind. The mind can afterwards go to various places along the walk in the mental model. The mind can float through the walk in the mental model from beginning to end, and if the memory serves well, the mind can float back in the reverse direction from the end of the walk to the beginning. Floating about in a mental model is a standard activity of the mind.

Reality is an ongoing interrelation of situation developments, and it is also hierarchically organized from elementary particles to galactic clusters. To use the mental model of reality as a tool for the organization of knowledge and for exploration and discovery of new knowledge, the mind can develop the practice of floating forwards and backwards along the structural logic based situation developments, and along the ongoing interrelations of those simultaneous diverse developments, and also up and down the hierarchic organization of reality.

Because reality is hierarchically organized, and because a large proportion of the factors and relations that occur at the various levels are either unique to their levels or have unique forms at different levels, science has become hierarchically fragmented. The different disciplines tend to study features of particular levels. Physics holds sway over elementary particles and atoms. Chemistry investigates the levels dominated by molecular relations. Biology examines life. Psychology deals with mind. And astronomy studies the nature of celestial bodies and their interrelations.

These are not hard and fast territories. There is, especially in recent decades, much crossover, for example the roles of elementary particles and atoms in astronomy, and molecules in biology. However—and despite the efforts of interdisciplinary programs—the hierarchic organization of reality, the vast quantity of knowledge now accumulated at each of the disciplinary levels, and the resulting educational traditions, have had the consequence that scientists have always been, and for the most part still are, trained as specialists. This is unavoidable. Science progresses day by day by way of the work of specialists. The content of the mental model is the consequence of their efforts.

Still, there is the now well known problem of specialists in the different disciplines having little or sometimes no significant understanding of science beyond their disciplinary boundaries. Specialists tend to have horizontally oriented world views centered on the levels of their disciplines. Metaphor is commonly used to transcend this limitation of viewpoint, but the results are questionable at best. Often factors of higher levels are attributed to lower levels. A comical example is the attribution of intention to unicellular organisms, and then there is the use of purposeful language in the description of the molecular biology of the cell. Inappropriate attribution of factors to stages of development and to hierarchical levels where they do not exist and do not play roles, and inaccurate use of descriptive language, make it easier to write, to talk, to communicate, but they mask an underlying failure to achieve a truly accurate understanding.

It is therefore, a significant feature of the mental model that it provides a world view with a vertical orientation based on the development of the hierarchical organization of reality. Factor development, situation development, and the interrelation of developing situations, make it possible to understand whether or not a factor can occur at some particular stage of development or level of organization, and if it can, what form it would have.

The mental model is a model of reality, all of it. Reality, however, is infinite and eternal, with an infinity of situation developments—every atom with its individual history is a situation development. Even though the model can contain only a sampling of this vastness, only the factors, situations, developments, and ongoing interrelations of developments of particular significance or interest, the model is going to be much larger than can be consciously viewed at one time. Just as when mentally walking through a familiar building the mind pictures only the part of current interest, when floating about in the mental model of reality the mind needs to image only the part that is currently being investigated.

The goal of the accurate analysis of the intrinsic nature of reality is the understanding of reality. The mental model is a summary of the current stage of that developing understanding. Each component of the model has a reality referent. It is the understanding of those reality referents that is the goal—for example, what is the origin and nature of emergence as a component of reality? The goal is not an understanding of the concept of emergence, but rather emergence itself. Let the intrinsic nature of reality dictate the nature of the understanding of reality. The nature of emergence must dictate the nature of the concept of emergence.

The orientation of the mind must always be toward those reality referents. Do not think of this book as a work of abstract philosophy. Do not think of the factors as abstractions. They are not Platonic forms, which do not exist in reality. Nor are they abstract concepts. A factor itself is to be used to construct the conceptual model whenever the factor can occur within the brain as itself, in its own intrinsic form. When some aspect-of-the-being of a factor precludes its occurrence within the brain in its intrinsic form, then a concept of the factor is used instead. The form of the concept must be based on and refer to the intrinsic nature of the factor. And the discussion is then not about the concept but about the reality-referent of the concept. The factors dealt with here are not abstractions or concepts. They are the factors of existence and organization that are-there, that actually compose reality.

Additional Notes on Method
The Mental Model Is Not a Theory

The mental model is a summary of the current stage of a continuously developing understanding. The goal is the understanding of the intrinsic nature of reality, here the intrinsic nature of emergence, and additionally of the foundational origins of cause and self-organization. For the sake of accuracy the mental model should not contain speculative components such as postulates or hypotheses.

A theory on the other hand is a tool of exploration. It is a combination of a summary of what is known plus speculation on what might be a fuller or more accurate understanding of the subject of the theory. A theory contains postulates and hypotheses, components which are not known to be true. These speculative components can be derived from structural logic, the nature of what is known providing hints at what might be known, or they can be derived from intuition or imagination and have no evident basis in reality. Theory, with its speculative components, has an intentionally much greater range of possibility as a tool of exploration. Theory is free to imply experiments to test the unlikely or even the apparently impossible. Both the mental model and theory can be used in a similar manner as tools for exploration, however the mental model is not designed for such use, while theory is. The utility of the mental model as a tool for exploration is an added benefit derived from the nature of the model.

Reduction

Constitutive reduction is a technique that effectively clarifies the nature of situations such as hierarchic organization and complex processes. With this technique a situation is dismantled into its separate components. This identifies the components and exposes their individual natures. Because the nature of the original situation was the consequence of the intrinsic natures of the components and their manner of coming together, and because the natures of the components determine for the most part the possible ways in which the components can interrelate, constitutive reduction becomes significantly more relevant when combined with synthesis. With the information gained by constitutive reduction, and with the use of structural logic, the original situation can be recreated, with observations along the way revealing the manner in which each component and its intrinsic qualities play their roles.

Because reality is fundamentally developmental and hierarchic in organization, constitutive reduction is necessarily a staged procedure, progressing stage by stage and level by level. This is unavoidable because of the role of emergence. While everything, in that part of material reality with which humans are familiar, is composed of elementary particles, the stages of development and the levels of hierarchic structure are patterns of organization. These patterns of organization are emergent, their existence and nature being dependent primarily on the immediately prior stage or level, that is they are dependent on the nature of the components of the prior stage or level and their manner of combining. The elementary particles are there and playing their required roles, but the intrinsic patterns of individual particles play roles only in the next stages or levels, and it is the emergent patterns of organization that play the dominant roles in further developments and higher levels. Constitutive reduction is thus a procedure of examining the pattern of organization of a level to identify the subpatterns of which it is composed, then examining each of those subpatterns to identify their component subpatterns, stage by stage, level by level, all the way down to the elementary particles.

This then explains the failure of explanatory reduction, which "...claims that all the phenomena and processes at higher hierarchic levels can be explained in terms of the actions and interactions of the components at the lowest hierarchic level" (Mayr, 1988: 11). By ignoring the roles of the emergent patterns of organization of the staged developmental organization and hierarchic level organization of reality, the goal of this form of explanatory reduction becomes physically impossible. The qualities and interrelations of the elementary particles are explanatorily destitute when it comes to elucidating the organization of a coyote skull, its component bones and teeth, and its shape.

To explain a coyote skull requires the analysis of all levels from the molecular to the ecosystem. The shape of a molecule is a consequence of factors of elementary particles and atoms. The roles of molecules in bone and teeth that are consequences of molecular shape, however, occur at a level of organization beyond that of their component particles and atoms. It is not the shapes of the particles and atoms that play these roles, but rather the shapes of the molecules. The shapes, or patterns of organization, of molecules again play roles in the DNA and proteins that play roles in the processes that create bone and deposit it in the distinct pattern of individual bones that comprise the skull. And that particular set of bones has an evolutionary history that goes back about 360 million years prior to the existence of coyotes, all the way back to the fish Eusthenopteron and its relatives (Alexander, 1994: 197). The process of change from the set of bones of that ancient ancestor to those of a currently living canine involves an immense number of physiological, ontological, and biological evolutionary relations which are above the levels where elementary particles play their specific roles. Finally, that evolutionary process occurred in relation to the ecosystems in which the coyote and all its ancestors lived, ate, and reproduced.

The understanding derived from constitutive reduction and synthesis is essentially an explanation of the origin and nature of a situation or level of organization. There could be, then, a form of explanatory reduction that explains by stages. Note that in this form explanatory reduction does not eliminate the need for a separate description or treatment of an advanced stage or higher level. It simply explains it in terms of its components and their togetherness.

Prediction

Just as the advanced stages and higher levels cannot be described or explained by way of the qualities of the first stages or lowest levels, the advanced stages and higher levels cannot be predicted from first stages or lowest levels. However, since each stage or level is the consequence of the togetherness of its components, it is possible to figure out a stage or level by examining the qualities of its components to see the various ways in which they can fit together. Since reality is the staged development of hierarchic levels, the organization of later stages and higher levels can be figured out by way of a staged process of analysis.

Emergence is the reason higher levels cannot be predicted from levels that are more than a stage or two below. It is not possible to predict straight from A to Z. However, emergence, followed by way of structural logic and development, can be a tool of prediction. The process must go emergent step by emergent step. In seeking understanding from what goes before to what follows, emergence is the bridge from one stage or level to the next.

Something

The term *something* refers not only to material objects or systems, but also to any factor of existence or organization.

Role

The term *role*, and the phrase *the role it plays*, are used because they seem to carry less anthropomorphic, teleological, function, and design baggage than other similar terms. The role something plays is simply the influence its existence has on a situation.

Defining

There are two basic types of defining. One is creative in that to varying degree the definition results from decision, for example, deciding on the rules for a game, or deciding what a term will mean in some special context, such as law.

This form of defining plays an important role in the social and cultural context. The other is defining by description. This is both a process describing something that exits, and of differentiation between that something and other things that exist. It is a matter of recognition rather than decision. It follows the prime imperative of the accurate analysis of the intrinsic nature of reality, Look to reality itself. Let the nature of reality dictate the nature the understanding of reality. Let the nature of what is being defined by way of description dictate the content of its description and the manner in which it is differentiated from other things.

Order of Presentation

Reality is hyper-relational in that each occurrence of each factor has multiple relations with the other factors with which it is coexistent or sequentially organized. Language is sequential. In description, the networks of factor relations must be teased apart and the descriptions of their components strung end on end as a chain of sub-descriptions. It is left to the reader to put them back together again in the nonlanguage, nonsymbolic, mental model. To facilitate this process of reassembly, the chain of sub-descriptions is given a specific order of presentation. This follows, as much as is possible, the natural order that exists among the networks of factor relations. The order of presentation follows developments from prior to later, from what goes before to what comes after, and follows hierarchy from lower to higher, thus following pathways of existential-dependence wherein the existence of one factor is dependent on the existence of another factor. Generally, this follows the path from the simple to the more complex, which works out well because earlier stages and lower levels, being simpler, are easier to understand. The components can be slowly reassembled, allowing the complexity of the understanding to build in parallel to the increase of complexity intrinsic to the situation under investigation.

Sometimes factors are presented out of order, or in different order in different parts of the book. There are various reasons for this. Sometimes with the explanatory mode, a situation can be best clarified by following relations other than from the simpler to the more complex or from earlier stages to later stages. With some factors the initial explanation can be made more clear if it is delayed and then presented within the context of other factors. When factors are coexistent, it sometimes does not matter whether the description of the situation begins with the one factor or the other.

History of the Developing Understanding of Emergence
This book is about the origin of emergence as an aspect of the intrinsic nature of reality. It is not about the concept of emergence or the history of that concept.

Errors and Contrary Opinion about Emergence Are Not Discussed
For the purposes of brevity and clarity, this book is written in a descriptive form somewhat like a biologist does when describing a new species. It is, for the most part, just a description of what is observed, with additional material derived from what is observed.

Deeper Origins—Foundations, Developments, Interrelations, and Origins That Are Basic to the Origin of Emergence

To understand why emergence occurs, to understand why and how something new can come into existence, it is necessary to explore deeper into the nature of existence and organization than the level of the origin of emergence itself. To understand the nature of emergence, it is necessary to look deeper into the origins of the various factors of existence and organization that together constitute the process of emergence in order to see why they have the consequence that emergence occurs.

Chapter 1: Introduction

Emergence is the universal determinate process of creative change based on consequent-existence by which newly occurring patterns of material organization come into existence. There are several factors here, (1) the determinate aspect of reality, (2) consequent-existence, (3) change, (4) newness, and (5) organization, that play roles in the origin and intrinsic nature of emergence, and all originate prior to the origin of emergence. To understand emergence it is necessary to understand the foundation and initial development of these prior factors. The following three chapters will follow them from their foundations, up through their developments and interrelations, to the origin of emergence. Along the way will be found several other factors such as continuance, sequentiality, combinatorial enhancement, sequential enhancement, and initiators whose origins, roles, and developments are of significance.

Factors and developments that originate or occur simultaneously will be presented sequentially, beginning in each case with those factors or developments that are more simple in character, followed by those with progressively more complex natures. Here and there are comments about the significance of various factors for the origin and nature of emergence.

Reality

Only that which is real can play roles in the existence and intrinsic nature of emergence.

Reality is that which exists. There is but one reality—all that exists. Existence is the ultimate foundation. As will be explained more fully in the chapter on space, the difference between existence and nonexistence is complete. Existence and nonexistence are distinct, without any condition, state, or form of partial being between the one and the other. It is an either/or situation. Whatever exists, exists. Whatever it is, that is what it is. By way of its existence, that which exists is itself. By way of existence, it has intrinsic self-identity.

There are two known primal-forms-of-existence that together constitute reality—space and the primal form of matter. As the foundation of reality, they constitute the existential foundation of all that exists. Motion and energy are so existentially-dependent on matter that they are herein treated as matter based factors. Each of the two primal-forms-of-existence has its own unique mode-of-being, or manner of existing, space as immaterial extensional place, and primal matter with its substantial manner of existence. Each plays roles with emergence according to the intrinsic qualities of its particular mode-of-being.

Space is the foundation of reality.

Chapter 2
Space

Space provides the foundations of place, extension, continuance, quantity, sequential-difference, parts, coexistence, and determinate-reality, and it provides the origin of initiation, consequent-existence, continuance-of-being, change, universal newness, and development. Space also provides an existential context, a place-to-be, for all that exists. To exist at all, every aspect of the being of that which exists must conform to the existential context that is space. Space provides the place wherein emergence occurs and to which all aspects of emergence conform.

Space Exists—Space Is Place
Spatial Place Is Immaterial

Because space is immaterial, it can have only those characteristics or qualities appropriate to an immaterial mode-of-being. That which is immaterial cannot have variation in the nature of its mode-of-being because there is nothing existing there with the capacity to be different in mode-of-being. Space is uniform. That which is immaterial cannot have motion as an aspect of its mode-of-being because there is nothing existing there to move. Space is static. That which is immaterial cannot interact with other modes-of-being because there is not anything there with the intrinsic capacity to interact. Space cannot affect matter, and matter does not have the capacity to affect immaterial space. Space is the only immaterial aspect of reality. The immaterial aspect of reality does not develop.

Spatial Place Has Extension

For space to exist as place, there must exist some quantity of place, some foundational existential quantity of place, for to have no existential quantity at all is to not exist at all. The existential quantity of space occurs as the extension of spatial place and is thereby a primal aspect of reality. Spatial place is immaterial—spatial existential quantity is immaterial. The quantity of spatial extension is the first stage of the development of quantity. Quantity, in its initial mode-of-being, is immaterial and therefore without units. The primal foundation of quantity is nonnumerical. Quantity in its primal pre-developmental stage as the existential quantity of spatial extension is a qualitative aspect of reality.

The extension of spatial place is continuous. For existential quantity to exist, it must exist in a constant, uninterrupted, continuous manner. Because space is immaterial, the extension of spatial place is uniformly continuous. The continuous nature of spatial extension is a primal aspect of reality.

The continuous extension of spatial place is voluminal, as summarized with three-dimensionality. To not be voluminal, to not be three-dimensional, is to not exist at all. To not have extension in one dimension is to not have anything existent there to have extension in the other dimensions. Spatial voluminal extension is requisite for existence. (There are three aspects of space, three types of spatial location, point location, line location, and plane location that are dimensionally limited. The modes-of-being of these spatial locations are, however, existentially-dependent in that they are dependent for their existence on the voluminal extension, the voluminal existential quantity, of the surrounding spatial place.) Spatial voluminal extension is a primal aspect of reality, an existentially foundational aspect of reality. Existential quantity of any form is voluminal, spatially three-dimensional. That which exists, everything that exists, has spatial voluminal extension as an aspect of its mode-of-being.

The continuous voluminal extension of spatial place is infinite. The voluminal extension of immaterial spatial place is infinite because space cannot not exist.

Space cannot not exist because it is not possible to limit the immaterial extension of space. Because space is immaterial nothing can interact with space to limit it. Beyond any point in space is more space. Spatial place is without limit. Because space cannot not exist, all the continuous infinite voluminal extension of spatial place is coexistent. Space exists as an infinite immaterial continuum of place.

Spatial Place Has Parts

The extension of spatial place has parts. With nearly anything that exists, there is the part that is one half of its continuous existential quantity, and there is another existentially distinct part that is the other half of that quantity. Space is infinite, and therefore has no center. There do not exist two fundamental halves of the existential quantity of space. But any point in space can serve as a location, with all the infinite extension of existential quantity on the one side and all the infinite extension of existential quantity on the other. There is the part that is all the spatial place on the one side, and there is another existentially distinct part that is all the spatial place on the other side. The existence of parts of the continuous voluminal extension of spatial place is a primal aspect of reality.

These two regions of space, these two parts of space, are not fully infinite in the manner that is the entirety of space. There is an aspect of limit to each of these two infinite parts. They are infinite only in the direction of the spatial extension away from the point, or spatial planar location, of their separation. The extension of the individual parts ends at the plane of separation of their individual partness. They are extensionally limited in that manner.

The parts of infinite spatial extension occur as extensionally limited spatial places. All space is coexistent—all spatial places are coexistent. Because space is static, any particular spatial place is locationally distinct, a unique location. To be coexistent and locationally distinct is to be existentially distinct. Any particular spatial place is what it is, it is itself, individually unique and distinct from any other spatial place. Any particular spatial place has unique self-identity.

Relation

The parts of space exist relative to one another. They do so with specific distance and direction relations. All the parts of infinite space exist relative to one another in this manner because they are coexistent and because space is static and uniform. The existence relative to one another of extensionally limited parts of space, spatial places, is a primal aspect of reality, an existentially foundational aspect of reality. The coexistence of spatial places is the primal stage of development of relativity, of relation, and is foundational to all spatially locational and interactive relations of that which exists.

The coexistence of the parts of space is the primal stage of the development of coexistence. Infinite spatial extension cannot not-exist, and the parts thereof are thereby coexistent. The parts of spatial extension have coexistent, locationally distinct, individually unique self-identity. Because spatial extension is immaterial, static, and uniform, there exists between the coexistent parts quantitatively specific relations of distance and direction.

One way of saying two places are a certain distance apart is to say they are a certain spatial place apart. The nonadjacent spatial places are the primary components of a coexistence situation. The third spatial place that exists between them plays the role of the spatial relation between the two. The reality referent of the concepts of the distance and direction between two spatial places is a third spatial place of specific extension and positional orientation. That third spatial place is extrinsic to the primary components, and the spatial relation between the two is thereby also extrinsic to them.

In a coexistence situation there are primary coexistence factors and secondary coexistence factors. Primary coexistence factors play roles in every situation of coexistence. Examples of primary coexistence factors are (a) the state of coexistence itself, (b) the existence of each individual primary component of a coexistence situation, (c) the nature of each component, and (d) the simultaneity of their existence. Secondary coexistence factors play relational roles that occur extrinsically between the primary components, such as the specific positional orientation and extension of the spatial place between two other spatial places, the direction and distance relations. The nature of the secondary coexistence factors depends on the individual nature of each primary component and its location in space.

Combinatorial Enhancement

The coexistence of the parts of space is the foundation of combinatorial enhancement. If there is coexistence, there is relation. The relation is in addition to and existentially-dependent on the existence of the primary components of the coexistence. The relation is an enhancement of the situation. With the coexistence of spatial places the enhancements are the direction, distance, and sequential-difference relations between those places. Combinatorial enhancement is the occurrence of relations between components of a coexistence situation. In coexistence situations the primary and secondary coexistence factors are coexistent. This aspect of coexistence is the foundation of the developmental factor that the whole is greater than the sum of the parts.

A simple numerical summation of the components of a group does not include the relations between those components. In the coexistence situation of two separate spatial places, there are the two primary components, the two places, and there are the direction and distance relations between those components. That is a minimum of four factors in a coexistence situation of two components. In a situation that is nearly as simple as it is possible to be, the whole is greater than the sum of the parts.

But combinatorial enhancement has an even more powerful effect. In the coexistence situation of three separate spatial places, there are the three primary components, and there are the direction and distance relations between them. But in this case each primary component has relations with two other components. There are three components, three direction relations, and three distance relations, for a minimum of nine factors in a coexistence situation of three components. Even with just spatial relations, distance and direction relations, the sum of the relations increases faster with additional components than does the sum of the components. Combinatorial enhancement develops interrelationally here with sequential enhancement, another form of enhancement which will be encountered a little further along. In this example sequential enhancement is the sequential increase of additional components.

Combinatorial enhancement has a major role in the origin of complexity. The whole consists of both the primary coexistence factors and the secondary coexistence factors, that is, the whole is the sum of the primary components plus the relations between them. The nature of the whole is a consequence of the coexistence of the natures of the primary components and the natures of the relations between those components. Because it is an aspect of spatial extension, an aspect of the relations between the parts of space, the foundation of combinatorial enhancement is a primal aspect of reality. Reality is complex because one of the major factors in the origin of that complexity, the factor that makes the whole greater than the sum of the parts, is foundational to all reality.

Sequentiality—Sequential-Difference

The parts of the continuous extension of spatial place exist sequentially relative to one another. Adjacent to any particular part of space there exists another individually distinct part of space that has its own unique self-identity. And beyond that part there exists yet another individually unique part of space. Throughout voluminal spatial extension there exists uniformly continuous sequential-difference of spatial place self-identity. The sequentiality of spatial extension is immaterial, static, uniform, and continuous without structurally based units.

The sequentiality of space, voluminal extensional sequential-difference, is the foundational level of sequentiality as a factor of reality. It is a primal aspect of reality, an existentially foundational aspect of reality. Spatial sequential-difference is foundational for the existence of other forms of extensional sequential-difference such as the linear sequential-difference of beads on a string or the planar sequential-difference of a checkerboard or a varied topography. As spatial sequential-difference is voluminal, it is foundational for voluminal forms of sequential-difference such as a layered onion or a planet like the earth with sequential-differences from its core to its crust.

Voluminal sequential-difference does not require a central point from which the difference radiates out in every direction. In fact, voluminal spatial sequential-difference is best understood as a volume of sequential-difference. The understanding can be approached by way of a material voluminal mass of sequential-difference. A block of crystal, glass, iron, or any continuous medium of matter is a voluminal mass of sequential-difference. Throughout the entire volume of the block of matter there occurs sequential-difference. With such a block of matter, the subunits, the molecules, atoms, and elementary particles, constitute the sequential-difference three-dimensionally. Space is immaterial and without structural subunits. But space has static, three-dimensional, existentially distinct, extensionally limited places. With space, the extensionally limited parts coexisting relative to one another constitute an organizational level of three-dimensional sequential-difference. But because space is immaterial and without units like atoms, the voluminal sequential-difference is actually uniformly continuous. Space is a volume of sequential-difference.

Extensional Development—Sequential Enhancement

Reality is a progressively ongoing interrelation of an apparently limitless variety of factors. Several of these factors, development, emergence, hierarchy, and evolution, can be used as tools of exploration and understanding, as maps or way-finders. There are three primary developmental pathways that originate prior to emergence and play roles in its origin. The first of these is extensional development, the development of spatial relations.

Extensional development occurs here in the form of sequential enhancement, sequential quantitative increase. From any spatial point location, spatial place is spread out in ever-greater extension. Because space is immaterial, this sequential quantitative increase, this sequential enhancement, is continuous, without units, nonnumerical. This primitive form of development is the primal foundation of sequential enhancement. With increased extension there is a corresponding increase of sequential-difference—the more extension, the more sequential-difference. Because space is infinite, the quantity of sequential-difference is infinite. From a point in space to infinite space there is a progressive increase in the quantity of sequential-difference. This also is sequential enhancement.

The coexistent parts of space have adjacent or nonadjacent relations with one another. Adjacent parts of space have direction relations, but no distance relations. Nonadjacent parts of space have both direction and distance relations. The distinction between adjacent and nonadjacent spatial places is the occurrence of extension between the nonadjacent places. This difference between the two situations is a case of extensional development. Extension plays a role in the second case, with the distance between nonadjacent places being an aspect of that role.

Because extension is required for there to be existence, extension is always there and it always plays a role. If there is existence, there is extension, and if there is extension, there is distance between the nonadjacent parts of that extension. Spatial extension, however, is continuous, and every part of spatial extension is surrounded by adjacent part of that extension. That is the first aspect or stage of extensional development—adjacent directional relation. And the second aspect or stage is the existence and role of extension—there is nonadjacent distance relation.

The distinction between adjacent and nonadjacent primary components of a coexistence begins in extreme simplicity with spatial places. But with developed coexistence situations adjacent or nonadjacent can be of considerable importance. For example, cause, physical contact interaction between coexistent substantial factors, requires the adjacent relation, but extension, and the nonadjacent relation, play roles in patterns of nonadjacent coexistent units and groups, such as the letters and/or words in this sentence which are recognized and used by sentient beings.

Spatial sequential-difference is not a form of change. Accurate understanding of the intrinsic nature of reality requires a precision here in the understanding of the intrinsic nature of change, and in the definition of the term change, that does not follow common usage where the term is applied to both forms of sequential-difference. Sequential-difference plays roles in both the extensional developmental pathway and in the developmental pathway that results from the occurrence of change, change development. But the difference in the natures of the two cases is so profoundly fundamental that to apply the term to both extensional development and change development does nothing more than obscure this critically important distinction.

While sequential-difference plays a role in the nature of change, not all sequential-difference has an intrinsic aspect of actual change. Observe a ruler. The sequential-difference is marked off from one end to the other. As the ruler is left to sit idle on a table top, there exists there sequential-difference, but there is no change at the relevant level of observation. Any particular marked off section does not change its nature, say from that of a wooden ruler to that of a steel ruler or from the orderly arrangement of atoms and molecules of wood to that of a disorderly pattern of carbon, calcium, and phosphorus atoms. Nor does one section move over and occupy or merge with the marked off section next to it. There is no change in the intrinsic identity of the various parts of the sequential-difference, nor is there any change in their relative positions. Different parts of the ruler have different individual self-identities and they are sequentially organized, there is sequential-difference, but no change.

Change and its developments-of-origin will be examined further along. It will suffice here to understand that the distinction between these two forms of sequential-difference is the difference between coexistent sequentiality and noncoexistent sequentiality. As the ruler sits there on the table, all the parts of its sequential-difference are there together, simultaneously. They are coexistent.

Consider a marine snail shell of the type that is a long narrow spiral. There is a sequential-difference marked off from the large end, where the snail puts out its mouth to eat and its foot to crawl about, to the small more pointed end that was created by the snail when it was young. The sequential-difference is marked off by the series of spirals of the shell. As the shell is carried about the ocean floor by the snail, all the parts of the shell's sequential-difference are there together, simultaneously. They are coexistent.

After the snail dies and the abandoned shell has been covered by the accumulation of sediments, a process of fossilization can take place. The original matter of which the shell was composed is slowly replaced by other matter. The process preserves the pattern of the sequential-difference of the series of spirals. All the parts of the shell's sequential-difference are still there together, simultaneously, coexistent. But the material identity of the shell has changed.

With the fossilized shell the coexistent sequentiality first constructed by the snail is still there. But there has been a noncoexistent sequential change from the material the snail used in creating the shell to the replacement material deposited by a geologic process. The coexistent sequentiality has endured through the noncoexistent sequentiality of the transformation from the one material to the other.

While common usage labels the sequential-difference of a ruler as change, the difference between that sequential-difference and that of change is significant. Motion, emergence, and cause are all interrelational developments of these two distinct pathways of development, extensional development and the development of change, and it is not possible to understand the three without understanding this distinction.

Spatial Place, Existential Organization, and the Foundation of Pattern

The immaterial continuous voluminal extension of spatial place has organization in that the coexistent extensionally limited places of space exist relative to one another within the larger context of the infinite spatial continuum. The places of spatial extension exist sequentially relative to one another. Any particular place in static space is a specific distance and/or direction from any other particular place.

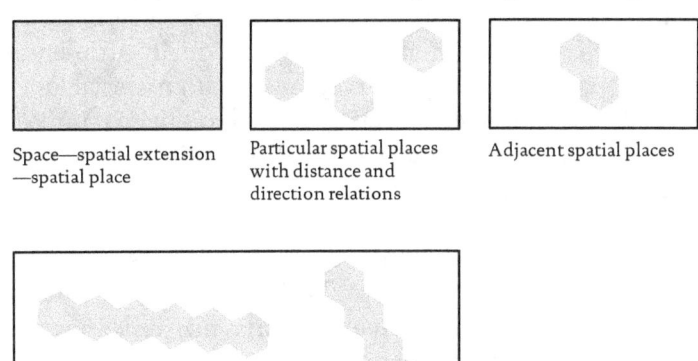

Space—spatial extension —spatial place

Particular spatial places with distance and direction relations

Adjacent spatial places

Sequential spatial places

Figure 2.1 *Immaterial pattern of extensional relations—Level one extensional pattern.*

Existence has organization, existential organization. The parts of a continuous existential quantity exist relative to one another—they exist in organizational relation to one another. Relative existence is a primal aspect of reality. Existential organization is a primal aspect of reality. Merely to exist is to have organization, existential organization.

The organizational factors of spatial place constitute the foundation of pattern. Spatial places have individual distinct self-identity based on their existence. They are also locationally distinct. The organizational factors of direction, distance, and sequentiality between static, coexistent, locationally distinct spatial places constitutes a pattern of relations—immaterial pattern differentiated locationally and by the distinct self-identity of the spatial places (Figure 2.1).

Extensional pattern requires coexistence of primary components with extensional relations, direction and usually distance. If there is coexistence, there is relation. The relation is in addition to and existentially-dependent on the existence of the primary components of the coexistence. This is combinatorial enhancement. Pattern is a combinatorial enhancement that involves the entire situation, the primary components and their extensional relations with one another. Immaterial locationally differentiated pattern of spatial place plays roles with material locationally differentiated pattern (internal to a primal unit of matter) and with material pattern differentiated by two or more units of matter in space.

Existential Context—A Place-to-Be

Extensional pattern develops, both by stages and by levels. The factor of levels develops when matter is there to play a role. Here, the extensional pattern of coexistent spatial places, first level pattern, has two stages of development, adjacent spatial places and nonadjacent spatial places. With adjacent parts of space, it is a pattern of static, three-dimensional, direction relations. These places and their direction relations provide the extensional place-to-be for the equivalent materially differentiated extensional pattern. With nonadjacent parts of space, it is a pattern of both direction and distance relations. This immaterial pattern also provides the place-to-be for the equivalent materially differentiated pattern.

The infinite continuum of spatial place provides an existential context, a place-to-be, for all else that exists, for all else that could possibly exist, for to not exist in spatial place is to not exist at all. Everything that exists, exists in space. Spatial place is existentially foundational to all forms-of-existence, all modes-of-being.

The extension of spatial place provides an existential context for the extensional aspects of that which exists other than space. The existential quantity of spatial place provides a place-to-be for the existential quantity of all else that exists. The sequential-difference of spatial extension is a place-to-be for other modes of extensional sequential-difference. The voluminal aspect of space provides the place-to-be for the voluminal aspects of other forms-of-existence and modes-of-being. And extensional development provides an existential context for developments of structural relations. All structural relations of that which exists conform to the spatial relations of the spatial places they occupy.

Existential Requisites, and the Intrinsic Self-Identity of Space

Existential quantity is requisite for existence. To have no existential quantity is to not exist. Spatial voluminal extension is requisite for existence. To not be three-dimensional is to not exist. If either of these two factors does not play its existentially foundational role in any particular situation, then there is not anything existing there. The situation does not exist.

The distinction between having existential quantity and thereby existing, and not having existential quantity and not existing, is complete. The distinction between having voluminal extension and thereby existing, and not having voluminal extension and not existing, is complete. Either that which exists has these factors and is-there, or it does not have these factors and is-not-there. The difference between existence and nonexistence is complete.

Whatever exists, that is what exists.

Space exists, and by way of that existence space has intrinsic self-identity. By way of its existence space is what it is. The qualities of space exist—it is infinite immaterial voluminal place, has parts and sequential-difference, and has static distance and directional aspects of organization—and because these factors exist, they are what they are, each with its intrinsic self-identity.

Spatial Place Exists, and It Continues to Exist

That which exists must have some continuance of that existence, for to have no continuance of existence at all is to not exist at all. The distinction between having continuing-existence and thereby existing, and not having continuing-existence and not existing, is complete. If there is existence, there is continuance of existence.

To exist, space must have continuing-existence. Space is a primal-form-of-existence, and the continuance-of-being of space is a primal aspect of reality. Space is infinite. The occurrence of the continuing-existence of space is infinitely omnipresent.

Continuance, Quantity, Parts

Because it is continuance-of-being, the continuing-existence of space occurs in a constant, uninterrupted, continuous manner. The continuing-existence of space is a form of continuance. Because space exists, there exist two forms of continuance, the continuance of spatial extension and the continuance of the continuing-existence of space.

Continuing-existence has no intrinsic units. Like quantity at its foundational stage, immaterial quantity of spatial extension, this stage of the development of quantity, spatial continuance-of-being quantity, is without intrinsic units, is nonnumerical, and occurs as a qualitative aspect of reality. Because space exists, there exist two forms of quantity, the existential quantity of spatial extension and the continuance-of-being quantity of the continuing-existence of space.

The continuing-existence of space has parts. Any continuous quantity has parts, for example, the one half of the quantity and the other half. With the continuing-existence of space, there was the part that has already occurred and there is the part that is occurring right now.

The parts of spatial continuing-existence are noncoexistent. Only the currently existing part is a component of reality. The previously occurring part of the continuing-existence of space no longer exists, and the part of that continuing-existence that will occur has not yet done so. The parts of the continuing-existence of space are not coexistent. While the quantity of continuance-of-being is a pure nonunitized continuance, as is the quantity of spatial extension, the quantity of continuance-of-being, the quantity of the continuing-existence of space, does not constitute an existing continuum.

To be noncoexistent is to be existentially distinct. The parts of the continuing-existence of space are different from one another, individually unique. Any particular part of the continuing-existence of space has distinct self-identity. The continuing-existence of space has parts, noncoexistent, existentially distinct parts which do not form an existing continuum. With the existence of space there are parts of two different forms, the parts of spatial extension and the parts of the continuing-existence of space.

Sequentiality, Sequential-Difference, Noncoexistent-Sequential-Difference

The continuing-existence of space occurs sequentially. During the previously occurring part, the currently occurring part did not then exist, and as the current part now occurs, the previous part no longer exists. The parts of the continuing-existence of space occur sequentially—they are noncoexistently sequential. To be existentially distinct and noncoexistently sequential is to occur in a pattern of sequential-difference—noncoexistent-sequential-difference. There are with space two different forms of sequentiality, of sequential-difference, the coexistent sequentiality of spatial extension and the noncoexistent sequentiality of the continuing-existence of space.

The sequential-difference of the continuing-existence of space is the foundational stage of the type of sequentiality that is specifically requisite for there to be motion, emergence, cause, and more developed situations involving noncoexistent-sequential-difference such as weather and biological development. These two modes of sequential-difference originate at different stages of the development of reality, and each is the foundation of a distinct path of development, one of sequential factors based on spatial extension, such as rows of objects, and the other of sequential factors based on continuance-of-being, such as sequences of events or processes. The coexistent sequentiality of spatial extension is an aspect of extensional development and the noncoexistent sequentiality of spatial continuing-existence is an aspect of change development. Throughout the development of the factors of reality these two forms of sequentiality, these two forms of sequential-difference, are continuously, developmentally, interrelated. Both types are requisite for emergence.

Relative Aspects of Noncoexistent-Sequential-Difference

The parts of spatial continuing-existence are noncoexistently sequential. They occur after or before relative to one another. The relative after and before relations of the parts of the continuing-existence of space are existentially foundational to all noncoexistent after and before relations of all that exists. Because space exists, there exist two different forms of relativity, the relativity of coexistent locationally distinct spatial places and the relativity of sequential noncoexistently distinct parts of the continuing-existence of space.

There are aspects of space that are relative, and there are aspects that are nonrelative. The existence and self-identity of space are not relative. Even though the distinction between existence and nonexistence is complete, an either/or situation, that which exists does not exist relative to its own nonexistence. It just exists. With existence, nonexistence is-not-there. With existence, nonexistence does not play a role, not even a relative one.

Existence is the foundation of self-identity. The two forms of relativity that occur with space, that of the parts of spatial extension and that of the parts of spatial continuing-existence, are aspects of the self-identity of space. Thus, the nonrelative aspects of space are the foundation of the relative aspects. The relative features of space, and their intrinsic qualities, are existentially-dependent on the nonrelative features of space.

New Part, Change, Change Development

As space continues to exist, new part of that ongoing existence is thereby continuously occurring. The current part of spatial continuing-existence is noncoexistently different from the previous part of that continuing-existence. When the previous part was occurring, this current part did not then exist. It exists now, and is now newly existent. This is the primal foundation of newness, and it plays a role in the occurrence of all developed forms of newness.

The continuing-existence of space is the foundation of change. It is a form of continuance, and any form of continuance has parts. New part of continuing-existence is continuously occurring. The parts of continuing-existence are noncoexistent, occurring sequentially in relation to one another. Being noncoexistently sequential, the parts have unique self-identity—they are sequentially, existentially, different from one another. The continuing-existence of space, the continuous transition from the previous part to the current new part, occurs as noncoexistent-sequential-difference, a transition from one state of a situation to a different noncoexistent state. Noncoexistent-sequential-difference is change. Because space is infinite and coexistent, because the continuing-existence of space is infinitely omnipresent, this form of change is universal.

Spatial continuing-existence is the foundational stage of change development. A consequence of the continuing-existence of space is that there has occurred an ever-increasing quantity of that continuance. Ever-increasing quantity is sequential enhancement, and sequential enhancement is a factor of development. It was seen before as a factor of extensional development with spatial place. In that case the sequential enhancement involved coexistent part, and that coexistence provided a role for combinatorial enhancement. If there is coexistence, there is relation, which is enhancement of the situation. Since the case with extensional sequential enhancement has static coexistent parts, there is no role for change.

Spatial continuing-existence is change, with non-coexistent part, and therefore no role for combinatorial enhancement. This is a different stage of the development of sequential enhancement. Rather than coexistent sequential relation, it is noncoexistent sequential relation. With noncoexistent relations, the sequentiality has a more significant role. The noncoexistent sequential after and before relation sets the stage for consequent-existence, determinate-reality, initiation, motion, emergence, cause, and all developments of change beyond continuing-existence.

Change in an Aspect of Self-Identity

Spatial continuing-existence is the development-of-origin for change in an aspect of self-identity. During the change from one part of continuing-existence to a following noncoexistent part, the continuance is maintained but the parts change. There is a continuous change of which part is currently in existence. Because the parts of continuing-existence have distinct self-identity, and because there is a continuous change of which part is existing, there is a continuous change of which self-identity is existing, change from the self-identity of one part to the self-identity of the following part. Thus there is a maintenance of the continuance, a continuous maintenance of the self-identity of that which is continuing to exist, spatial place, while there is continuous change of the self-identity of the currently existing part of the continuance, of the currently existing part of the continuing-existence of spatial place. This constitutes a maintenance of self-identity during a change in self-identity—a maintenance of the self-identity of spatial extensional coexistent-sequential-difference during the continuous change of the self-identity of the current part of spatial continuing-existence noncoexistent-sequential-difference.

The stages of the development of change in an aspect of self-identity occur in several different forms. The first form is, Change from one individually distinct part of continuing-existence to the following noncoexistent individually distinct part. There are three cases of this form, of which change in self-identity of spatial continuing-existence is the first.

It is the change in the self-identity of the distinct parts of continuing-existence that makes it possible for there to be change in relation to the unchanging existential and extensional aspects of something that exists. The change of self-identity from part to sequentially occurring part of spatial noncoexistent-sequential-difference makes it possible for there to be change in relation to spatial coexistent-sequential-difference.

For there to be difference in self-identity, that difference must be noncoexistent. The role of the change and newness of spatial continuance-of-being is that it provides the foundational noncoexistence required for the other sources of change and newness, such as motion and emergence. Merely to exist, the intrinsic nature of those other forms of change must conform to the intrinsic nature of that foundational noncoexistent-sequential-difference of spatial continuing-existence. And that foundation of spatial continuing-existence provides the enabling context of sequential noncoexistent change in self-identity such that other forms of difference in self-identity can exist, such as those that occur with motion and emergence

The continuing-existence of space is the foundational stage of the development of the relation between maintenance of self-identity and change of an aspect of self-identity. The organization of material reality is constantly emerging, in part because of this foundational role of spatial continuing-existence. Emergence constitutes a stage in this development. By way of continuing-existence there occurs the continuously ongoing maintenance of self-identity simultaneous to the continuously ongoing occurrence of newness. Emergence operates by way of the maintenance of self-identity simultaneous to change and newness. With the process of emergence there are aspects of maintenance of self-identity concurrent to aspects of change of self-identity, and these features of emergence are existentially-dependent on the maintenance of self-identity and the change of self-identity that occurs with spatial continuing-existence.

Consequent-Existence and Determinate-Reality

Emergence is the universal determinate process of creative change based on consequent-existence by which newly occurring patterns of material organization come into existence. Several of these factors have their foundations in the primal continuing-existence of space. The first two are consequent-existence and the determinate aspect of reality.

Consequent-Existence

If space did not exist, it could not continue to exist. Space exists, and thereby it continues to exist. The continuing-existence of space is an existential consequent of the existence of space. Continuing-existence is a consequence of existence. This is consequent-existence, the existence of one factor of reality as a consequence of the existence of another factor.

A development-of-origin is a change through which a factor comes into existence. The factor can originate as a factor of the development itself, or it can originate as a direct consequence of the development. For example, consequent-existence becomes a factor of reality as an aspect of the transition from existence to continuing-existence, while continuing-existence is the consequence of existence, the consequence of the transition from existence to continuing-existence.

With continuing-existence that which comes after is a continuance of that which has gone before. It is a continuance of the existence and self-identity of what has gone before. The existence and self-identity of what follows is thus a consequence, by way of continuing-existence, of the existence and self-identity of the initial factor. This is the foundational form of consequent-existence—consequent-existence by way of continuing-existence.

This first stage of consequent-existence, the continuously occurring consequent-existence that occurs with the continuing-existence of space, is a primal aspect of reality that plays a foundational enabling role in all further developed stages of consequent-existence, for example motion, emergence, cause, and evolution. Emergence is a process of consequent-existence, a developed form. Emergence originates as a stage in the development of consequent-existence, as does biological evolution, which also originates as a stage in the development of emergence. Consequent-existence plays a role in emergence, and both consequent-existence and emergence play roles in biological evolution.

Determinate-Reality

Consequent-existence is the basis of the determinate aspect of reality. The existence of what goes before determines, by way of its own continuance, the existence of what follows. Existence determines existence.

Existence is the foundation of self-identity. That which exists is itself, with intrinsic self-identity. The self-identity of what goes before determines, by way of its own continuance, the consequent self-identity of what follows. What it is that goes before determines what it is that follows.

Spatial continuing-existence is the continuance-of-being of the existentially based intrinsic nature of space. Consequent-existence by way of spatial continuing-existence is determinate, and constitutes the primal foundation of the determinate aspect of reality, that existence determines existence. Everything that exists does so within and conforms to the place-to-be, the existential context, provided by space. Therefore, the continuing-existence of all that exists occurs with and conforms to the continuing-existence of space. Universally the existence and identity of what goes before determines, by way of the continuing-existence of intrinsic self-identity, the existence and identity of what follows. Reality is determinate because existence is the foundation of self-identity, and because existence determines existence by way of continuing-existence of self-identity.

That existence determines existence is the core of consequent-existence. It plays several roles in the process of emergence, and provides to that process an aspect of complete determination of what follows by what goes before. It is the core of the determinate aspect of reality—of continuance-of-being, motion, emergence, cause, and all change and newness from the level of primal being all the way up the development of the organization of reality to the most complex realms of creation and order.

That reality is determinate bears on the status of free will. Free will exists, it is real—as life, sentience, cognition, and love are emergents of a determinate-reality, so also is free will emergent. Functional free will is an evolved factor of reality. Free will requires a determinate-reality for it to be of any significance. It requires a consistently orderly sequence of events to be of any practical use. Free will decisions in a chaotic nondeterminate world would be pointless. The factors of the origin of emergence are factors of initial situations, beginnings, foundations. Free will is a stage in the development of emergence, a stage in the epistemological existential-pathway-development. But it is one of the most developed and complex stages of that pathway, and is also one of the most recently evolved stages of that pathway as it exists on this planet. Free will occurs at the opposite end of the developments that play roles in the origin of emergence, and is thus not within the subject matter of this book. Its origins will be mapped out and its intrinsic nature will be described in a later work.

Continuing-existence is a form of noncoexistent-sequential-difference, making it a form of development, change development. It is a form of noncoexistent sequential enhancement, which is a transition from one state of a situation to another involving some form of enhancement of the resultant state, in this case the occurrence of new part. Because what follows in this case of noncoexistent sequential enhancement is determined by what goes before, the change development that occurs with spatial continuing-existence is a case of determinate consequent-existence. Existential-pathway-development is the progressive difference or transformation of a situation. With continuing-existence it is the transition from one part of the continuance to another. Existential-pathway-development, at all stages of change development, always has a component of spatial continuing-existence, and because of this, it is always based on determinate consequent-existence.

Existential-Dependency

With the existential-pathway-development that occurs with change development, following stages are existentially-dependent on the existence of prior stages. This is different from the existential-pathway-development that occurs with extensional development, where the following stages are coexistently there without this existential-dependency relation.

The unchanging factors of extension and the ever changing factors of continuing-existence are foundational to two distinct developmental pathways of the factors of reality, extensional development and change development, pathways that progress from the primal level of existence all the way to the highest levels of organization, and throughout that development these pathways constantly interrelate. Existential-dependency develops at various stages of that interrelational pathway.

Uniformity and Unidirectionality of Continuing-Existence

The continuing-existence of space is a uniform change because it is a continuance-of-being. Something that exists is-there, and continues to be-there. There is no role in being-there for the factors of faster or slower. Something cannot exist faster or exist slower. The change of the continuing-existence of space is uniform because it is the continuing-existence, the continuance-of-being of space. Space exists, it is-there, and it continues to be-there. There is no role in the continuing-existence of space for the factors of faster or slower. Space cannot exist faster or slower.

The continuing-existence of space is a unidirectional change. It is unidirectional because it is a continuance of existence, a continuing to be-there. When something exists, it is-there, and it continues to be-there for as long as it exists. That there is existence determines that there is continuance of that existence. The relation is from existence to continuance. Continuance derived from existence is one way—unidirectional.

Continuing-Existence and Organization

With the continuing-existence of space there occurs another form of organization. The change that is the continuing-existence of space is organized in that it is sequential and unidirectional. That it is sequential and unidirectional has the consequence that the parts of the continuing-existence of space are organized in specific noncoexistent after and before relations with one another. That the change of the continuing-existence of space is uniform has the consequence that the parts of the continuance are organized in quantitatively specific after and before relations. Because space exists, and thereby continues to exist, there occur two different kinds of organization, that of extension and that of continuing-existence.

To exist is to have quantity. To have quantity is to have parts. To have parts is to have relative relations. To have relative relations is to have organization. Simply to exist is to have organization.

The organizational factors of spatial continuing-existence constitute the foundation of pattern based on noncoexistent-sequential-difference. The parts of spatial continuing-existence have individual distinct self-identity based on their existence and differentiated by their noncoexistence. The organizational factors of spatial continuing-existence, the factors of sequentiality, unidirectionality, and specific after and before relations between distinct parts, their noncoexistent-sequential-difference, constitutes a pattern of relations—noncoexistent sequentially differentiated pattern. This is a different form of pattern, with a different mode-of-being, from the pattern of spatial place. That first form of pattern was extensional pattern, coexistent-sequential-difference, while this form is the pattern of noncoexistent-sequential-difference.

Spatial Continuance-of-Being Has Always Occurred and Will Always Occur

The infinite immaterial continuum of spatial place cannot not exist. Space has always existed. The continuing-existence of space is without beginning. There has occurred an unlimited quantity, an infinite quantity, of spatial continuance-of-being. The continuing-existence of space results in an ever-increasing quantity of the ongoing existence of space that has occurred. There was no beginning. There will be no end.

Simultaneity

Because all space is coexistent, all individual parts of space are coexistent with all other parts. All parts of space continue to exist together, simultaneously. The continuously new current part of the continuing-existence of a spatial place occurs simultaneously with the continuously new current part of the continuing-existence of other spatial place. And the continuously changing noncoexistent-sequential-difference of the continuing-existence of a spatial place occurs simultaneously with the continuously changing noncoexistent-sequential-difference of the continuing-existence of other spatial place. The simultaneity of the continuing-existence of distinct spatial places provides the situation in which there can be simultaneous differences occurring at distinct places. For example, an event can occur at one spatial place but not at an adjacent place, or vice versa. Or events can occur simultaneously at different spatial locations. All events that are occurring at the current part of the continuing-existence of infinite coexistent space are simultaneous.

Spatial Continuing-Existence Existential-Context

Space provides two forms of existential context for that which exists other than space, one a place-to-be for its extension, and another for its continuing-existence. Spatial place provides a context for that which exists, for all that exists. The extension of that which exists other than space, its existential quantity, extensional continuance, three-dimensionality, its coexistent parts, and its organization exist within and existentially conform to the place-to-be provided by the extension, quantity, extensional continuance, three-dimensionality, parts, and organization of spatial place.

There are patterns of material organization, from molecules to mountains, that come into existence, continue to exist, and then through one process or another are disorganized and cease to exist. These patterns of organization have limited quantities of continuing-existence.

The existence of the material pattern occurs during a specific limited part of spatial continuing-existence, with quantitatively specific after and before relations with all the continuing-existence of space that has occurred or that will occur. The pattern occurs at specific durationally limited part of the ongoing, eternal, continuing-existence of space.

The continuing-existence of that which occupies space occurs with the continuing-existence of the occupied spatial place, and thereby with the continuing-existence of all space. The newness, the noncoexistent-sequential-difference—the change—that occurs with the continuing-existence of that which exists other than space occurs simultaneously with, and conforms to, the newness and noncoexistent-sequential-difference—the change—that occurs with the continuing-existence of space. The noncoexistent-sequential-difference of spatial continuing-existence provides an existential context for other forms of noncoexistent-sequential-difference such as that of material continuing-existence, motion, and the other forms of change development.

The unchanging factors of spatial place, and the ever changing factors of its continuing-existence, are the foundation of reality. This extraordinarily simple, immaterial situation provides the full primal context for all else that exists, for substantiality, material extension, and material continuing-existence, for organization and structure, for all forms of change, motion, cause, and function, for galaxies and all they contain.

The mode-of being of the continuing-existence of spatial place is different from the mode-of-being of the extensional aspect of spatial place. The factors that play roles in the nature of space, (for example extension, voluminality, existential quantity, continuance, and sequentiality), are either entirely different or different in various respects from the factors that play roles in the continuing-existence of space. Even when the same factor plays a role in both cases, (for example quantity, continuance, or sequentiality), it does so in a different mode in each case. There is a summary of these differences in Appendix 1.

The Primal Development from the Existence of Space to the Continuing-Existence of Space

This is an existential pathway, existentially-dependent, multifactor development that involves direct situation development and indirect factor developments. It is an existential-pathway-development because it is the consequent-existence transition of an individual situation. It is an individual situation because infinite spatial place is coexistent. All infinite space exists and thereby continues to exist—simultaneously. The development from the existence of space to the continuing-existence of space is a development of a single infinite situation.

It is an existentially-dependent development because with consequent-existence the existence of the consequent is dependent on the existence of the initial situation. The continuing-existence of space is existentially-dependent on the existence of space. It is a direct development because the continuing-existence of space is an immediate consequence of the existence of space, without any intervening factors.

It is a situation development because change and enhancement of the situation occur. It is change because it is the transition from the mode-of-being of coexistent, three-dimensional extension to the mode-of-being of noncoexistent, nondimensional, nonextensional continuance-of-being. It is change also because factors occur with the development that are not constituents of the initial situation. It is the development-of-origin for continuing-existence, consequent-existence, determinate-reality, change, newness, change development, and noncoexistence. The origin of these factors constitutes an enhancement of the situation.

It involves indirect factor development because the factors of quantity, continuance, parts, sequentiality, relativity, organization, and existential context play roles in both the initial condition and in the consequent, but they do so in different modes-of-being in the two stages, with the forms of the prior stage not playing roles in the forms of the developed stage. The first stages are aspects of spatial extension, while the second stages are aspects of the continuing-existence of spatial extension. The extension, three-dimensionality, and coexistence of the mode-of-being of the initial stage of these factors do not play formative roles in the nonextensional, nondimensional, noncoexistent mode-of-being of their consequent developed stage. Both the first stages and the second are existentially-dependent on spatial extension. However, the first stages are direct intrinsic aspects of the being of spatial extension, while the second stages are only indirectly so, as qualities of the continuing-existence. The modes in which the various factors occur in their first and second stages are distinct because the modes of existence and of continuing-existence are distinct. Each stage of a factor derives its mode of existence either from the mode of existence itself or from the mode of continuing-existence. The second stages do not derive their modes of existence from the modes of existence of the first stages, and are thus not direct developments of the first stages.

These indirect factor developments are existentially-dependent in that the developed forms of the factors are aspects of the continuing-existence of space and are thus aspects of the continuing-existence of the initial forms of these factors.

Enhancement occurs with the indirect development of these factors (quantity, continuance, parts, sequentiality, relativity, organization, and existential context), from their form in spatial place to their form in the continuing-existence of spatial place, in that with the consequent situation they each occur in a second, new, and different form. The consequent forms of these factors are developed forms because they are different from and existentially-dependent on the initial forms.

The Initiators

The mere existence of space initiates spatial continuing-existence. There is no role for cause. There is no role for energy. There is no role for motion. There is no role for matter. At this primal foundational stage of the development of the change aspect of reality, immaterial existence alone initiates continuing-existence.

- Existence initiates continuing-existence;
- Continuing-existence is existence, ongoing existence;
- Existence initiates ongoing existence, and;
- Existence initiates existence.

This is the development-of-origin for five factors, the determinate aspect of reality, consequent-existence, change, newness, and change development. Initiators are those factors whose mere existence results in the determinate consequent-existence by which change, newness, and change development occur.

Initiation and Intrinsic Consequence

The existence of space initiates the continuing-existence of space. This is consequent-existence, the existence of one factor as a consequence of the existence of another factor. In this case, it is consequent-existence by way of continuing-existence. It is immaterial voluminal spatial place continuing to exist as immaterial voluminal spatial place. The consequence is space continuing to be space. In a situation where a factor continues to exist as itself there are no roles for any factors other than that factor initiating its own continuing-existence. With this form of initiation, initiation by way of continuing-existence, the consequence is entirely intrinsic to the initiator.

Determinate Aspect of Initiation

The role of initiators is determinate. Initiation and the determinate aspect play their roles together, but there is a difference between those roles. Initiation plays a role in the existence of change in that existence initiates new existence, existence initiates change. The determinate aspect plays a role in the nature of change in that the nature (including the existence) of what goes before, the initiator, determines the nature (including the existence) of what follows, the consequence. Initiators initiate consequent-existence. Consequent-existence is determinate. If a particular initiator exists, a particular consequent will exist. Existence determines existence.

Development of Initiators

Like most other factors, initiators develop. The first stage is the initiation of spatial continuing-existence by the existence of space. It is eternal initiation by way of primal existence. When the existence of space initiates the continuing-existence of space, it does so continuously, such that there occurs continuously new part of spatial continuing-existence, new part with sequentially distinct new self-identity. Because this stage of initiation occurs continuously, static unchanging spatial place occurs at continuously new part of its continuing-existence, at constantly changing, sequentially distinct, new part of its continuing-existence.

The existence of space as an initiator is particularly significant in three ways. First is its role as an initiator. It is a primal-form existential initiator because it is a primal-form-of-existence and because it is one of three factors of reality, space, matter, and motion, known to have a completely intrinsic aspect of self-identity, and are thus the only three factors of reality known to initiate continuing-existence. Second, in its role as an existential initiator, the existence of space is the universal foundational initiator in that all change of all forms is existentially-dependent on its consequent, spatial continuing-existence, which provides the existential context to which all forms of change must conform in order to exist at all. The roles of all other initiators conform to the role of this universal foundational initiator. And third, again in its role as an existential initiator, spatial existence is one of the three factors that are the foundational origins of all change. Change from one part of spatial continuing-existence to another part of that continuance is one of the three sources of change from which all other types of change are developed.

There are a few other basic types of initiators and initiation situations that will be found further along. As stated in the short description of emergence, there are several sources of change and newness the combined interrelational development of which constitutes the process of emergence. The process of emergence itself, at its origin, will be found to be a type of initiation situation, organizational initiation. The form of initiation situation that is the origin of emergence provides the existential basis of a seemingly limitless development of organizational initiation. This is one of the reasons it is important to understand emergence as a core factor of the development of reality, and why emergence is important to understanding reality in general.

In the development of initiators the prior stages tend to play roles in the later stages, those later stages generally consisting of combinations, interrelational developments, of the prior stages, plus additional factors that give the stages their particular character. The process of emergence, as an initiator of change and newness, is an interrelational development of several prior initiators which must be understood as a prerequisite to understanding emergence.

Initiators and Sequential-Difference

There are twelve cases of sequential-difference that play roles in the origin of emergence. One of these is the static unchanging coexistent-sequential-difference of spatial place. The eleven others are cases of changing noncoexistent-sequential-difference. As cases of change, they are all associated with initiators, and codevelop with them. The first of these eleven is spatial continuing-existence, the continuing-existence of the existential context in which emergence takes place, and to which all aspects of emergence conform. A later case of noncoexistent-sequential-difference, associated with another initiator, plays a role that interrelates them all together in an ongoing manner such that there emerges change in material organization, which is the development-of-origin of emergence.

The Continuing-Existence of Space Provides the Temporal Aspect of Reality

Spatial continuance-of-being is the reality referent of the term time. The factors of the continuing-existence of space are the factors of time.

Space is immaterial. As an aspect of the existence of space, the continuance-of-being of space is also immaterial. Time is immaterial. Because space is immaterial it cannot interact with other forms-of-existence such as matter. Time, the continuing-existence of space, cannot interact with matter.

Space is infinite. The continuance-of being of space is infinitely omnipresent. Time is omnipresent.

Space exists and cannot not-exist. That which exists must have some continuance of that existence, for to have no continuance of existence at all is to not exist at all. Because spatial place cannot not-exist, the continuance-of-being of spatial place cannot not-occur. Time cannot not-occur.

The continuing-existence of space results in an ever-increasing quantity of the ongoing existence of space that has occurred. There has been an ever-increasing quantity of time that has occurred. The increasing quantity of the continuing-existence of space that has occurred is a continuance-of-being, and is thus a nonunitized quantity. The increasing quantity of time that has occurred is a continuance-of-being, and is, intrinsically, nonunitized.

Because space cannot not-exist, space has always existed and will always exist. The continuance-of-being of space has always occurred and will always occur. Time has always occurred and will always occur.

The continuing-existence of space is without beginning. There has occurred an unlimited quantity, an infinite quantity, of spatial continuance-of-being. Time is without beginning. There has occurred an unlimited quantity, an infinite quantity, an eternity of time.

Time is not eternal like space is infinite. A region of space limited by a planer location is extensionally infinite in the direction away from the planer location but is extensionally limited at that location, and in that manner that region of space is not fully or entirely infinite as is space as a whole. The situation with time is similar, but the limiting factor for time is complete in that the part beyond the limit does not exist. The part of the continuing-existence of space that has not yet occurred has not yet contributed to the ever-increasing quantity of the continuing-existence of space that has occurred. The future has not yet occurred and has not yet contributed to the time that has occurred. There has occurred an infinite quantity of the continuing-existence of space before the current existing part of that continuance, but the existing part is the current limit of the continuing-existence of space. The quantity of the continuing-existence of space is not fully infinite. There is an aspect of its existence that is limited. There has occurred an eternity of time before the present, but the present is the current limit of the continuing occurrence of time. The quantity of time is not fully eternal. The quantity of infinite space cannot increase, it is existentially infinite. The quantity of time is ever-increasing, it cannot be existentially eternal.

The continuing-existence of space results in an ever-increasing quantity of the existence of space that has occurred—and quantity has parts. Only the currently existing part of the continuing-existence of space is a component of reality. The previous part of the continuing-existence no longer exists, and the part of that continuance that will occur has not yet done so. The ever-increasing quantity of time has parts. Of time, only the present is a component of reality. The past has ceased to exist and the future has not yet occurred. The parts of the continuing-existence of space are noncoexistent and existentially distinct with individual self-identity. Thus, when the parts of spatial continuance-of-being are represented as time, the parts are noncoexistent and existentially distinct with individual self-identity.

All spatial place is coexistent. The initiation of the continuance-of-being of all infinite spatial place is coexistent, concurrent. The current part of spatial continuing-existence is omnipresent. The current part of time is omnipresent. The present is the same throughout infinite space.

While the quantity of continuance-of-being is a pure nonunitized continuance, as is the quantity of spatial extension, the quantity of the continuing-existence of space does not constitute an existing continuum. Time is a pure nonunitized continuance but does not constitute an existing continuum.

The individually unique noncoexistent parts of spatial continuance-of-being occur sequentially. Sequential-difference is a characteristic of the continuing-existence of space. The individually unique noncoexistent parts of time occur sequentially. Sequential-difference is a characteristic of time.

The continuing-existence of space is a form of change. The transition from one part of the continuing-existence of space to a following part is change. The transition from one part of time to the following part is change. The continuing-existence of space is a uniform unidirectional change in that it is a continuance-of-being. Time is a uniform unidirectional change in that it is the continuance-of-being of space.

While spatial place is extensional, three-dimensional, and coexistent, the continuing-existence of space, as a form of ongoing change, is not coexistent, nor extensional, nor dimensional. Time is noncoexistent, nonextensional, and nondimensional. The present, as a form of ongoing change, has no length, no extension, and no dimensional aspect. The existence of the present is dependent on the existence of space. The ongoing present is dependent on the continuing presence of what space is, on the continuing presence of place.

The change that is the continuing-existence of space results in the occurrence of newness. When the previous part was occurring, this current part did not then exist. It exists now, and is now newly existent. When the previous part of time was occurring, this current part did not then exist. It exists now, and is now newly existent. New time is the newly occurring part of the continuing-existence of space. The initiation of new part of spatial continuing-existence is continuous. The present is forever new. This continuous initiation of new part of continuing-existence is like a never ending beginning. Due to its mode of origin, time is forever beginning.

The continuing-existence of space has organization in that it is unidirectional and sequential, the parts occurring with specific quantitative after and before relations. Time has unidirectional and sequential organization with specific quantitative after and before relations in the occurrence of its parts.

The continuing-existence of space provides an existential context for the continuing-existence of that which exists other than space. Time is the temporal context for the ongoing existence of matter, objects, systems, events, and processes. Only the currently occurring part of the continuing-existence of space is a component of reality. To exist is to exist in the current part of the ongoing existence of space. Of time, only the present is a component of reality. To exist is to exist in the present; for to not exist in the present is to not exist at all.

Because the parts of spatial continuing-existence are not coexistent and are not parts of a continuum, nor of extensional place of any sort, they do not have distance and direction relations. There is no geometry to continuing-existence. There are no geometrical aspects to the intrinsic nature of time.

The continuing-existence of space is determinate because it is a case of consequent-existence wherein what goes before determines what follows by way of the continuing-existence of intrinsic self-identity. Time, as the continuing-existence of space, is determinate.

Time—In Summary

Time, as the continuing-existence of infinite space, is an omnipresent, immaterially noninteractive factor of reality. It occurs as a continuous, uniform, unidirectional, sequential change, which cannot not-occur, and was thus without beginning and will never end. There has occurred an unlimited quantity, an eternity of time. As a continuance-of-being, there continuously occurs new part of time, which results in an ever-increasing quantity of the temporal continuance that has occurred. Temporal quantity occurs as continuance-of-being, and is thus a nonunitized quantity. Time has organization in that it is sequential and unidirectional, the noncoexistent, existentially distinct, individually unique parts occurring with specific quantitative after and before relations with one another, giving time the characteristic of sequential-difference. The existing or current part of time, the present, is the same throughout infinite space. The present, existing as a continuance-of-being, as a form of continuous change, has no extension, no existing quantity, no length. Of time, only the present is a component of reality, the past having ceased to exist and the future not yet having come into being. Therefore time does not constitute an existing continuum. Time is the temporal context for the ongoing existence of matter, objects, systems, events, and processes. Everything that exists does so simultaneously in the present, in the current existing part of the continuing-existence of space.

Because the past no longer exists and the future has not yet occurred, that is, neither the past nor the future exist, time travel is impossible. There is no place to go.

Matter conforms to space.

Chapter 3
Space and Matter

Matter provides developments of initiators, continuance-of-being, consequent-existence, determinate-reality, change, newness, coexistence, relativity, organization, pattern, sequential enhancement, and combinatorial enhancement foundational to emergence. Additionally, matter provides motion and the substantial basis of emergent organization.

There is still something fundamentally awesome to be discovered about matter. We do not know what it is. We do not know why matter exists. We have not achieved an understanding of the intrinsic nature of the primal mode-of-being of matter, its substantial nature.

There is a problem investigating the primal mode of material existence. We do not know if any of the elementary particles represent primal forms of matter because we do not really know what they are. We do not know what they are in the direct manner of experience, aided by some careful thinking, as we can with space. Nor do we know what they are in the indirect manner provided by science, aided by technology and some careful thinking, as we can in the case with atoms. Although, at the time of this writing some progress is taking place with the internal structure of protons, for example. But even that is still more of the same type of analysis that brought us down to the level of elementary particles, the discovery of the subunits of a whole. That still does not provide an understanding of the primal existential identity of matter.

At the time of this writing there is no statement of fundamental identity for matter equivalent to, Space is place. The statement, Matter is substantiality, cannot serve because that is like saying, Space is immaterial. Immateriality is a characteristic of spatial place, and substantiality is a characteristic of matter. The question can be asked, What is it that is immaterial? And the answer is, spatial place. That leaves the unanswered question, What is it that is substantial?

It must be kept in mind that none of the levels or parts of matter to which we currently have access may be the primal or foundational form of matter, and there is the suspicion that there may be two or more primal-forms-of-existence that in combination constitute the forms of matter we know about. For the purpose of understanding emergence as it exists, it seems obvious then to pick up the development of the intrinsic substantial organization of matter at the level where current understanding and technology provides access, and that is for the most part the level of the various elementary particles.

However, at the elementary particle level of the organization of material reality, emergence is already a universal factor. To understand the deeper origins of the material factors that play roles in the origin of emergence, it is necessary to figure out certain aspects of the nature of matter that are prior to the level of elementary particles. These are certain aspects that must occur for there to be existence, and certain aspects of the primal mode of material organization associated with the manner in which matter exists in space. Because they are factors of existence and association with space, these aspects are universal, applying to any mode of material existence.

Using Space to Understand Matter

What is understood about the nature of space can provide some insight into these fundamental aspects of the nature of matter which are important for understanding why emergence occurs. Matter, like space, has two major foundational developmental stages, the existence of matter and the continuing-existence of matter. There are factors with each stage that are required if there is to be existence. Both space and matter have these factors as intrinsic aspects of their distinct modes-of-being.

Chapter 3: Space and Matter

With the first stage, that of existence, there are other factors which are unique either to space or to matter, infinite immaterial continuum and extensionally limited discontinuous substantiality for example. These are the qualities of the individual modes-of-being of space and matter, qualities of their distinct forms of existential quantity. Understanding of the roles of these factors with matter must be obtained by way of examination.

Fortunately, the second stage, that of continuing-existence, is identical for everything that exists or that could possibly exist, and understanding the continuing-existence of space provides the foundation for understanding the continuing-existence for everything.

One of the factors that makes what is known about space useful for understanding matter is that space provides an existential context to which the factors of matter must conform. Because the voluminal extension of space is all the immaterial place that exists in any particular location, and because space is infinite, spatial place is all the immaterial place that there is. There is no other place where matter could exist. To exist is to exist in space. To not exist in space is to not exist. The material or substantial mode-of-being of matter occupies the existential context that is the immaterial mode-of-being of spatial place, and the continuing-existence of matter occurs with the continuing-existence of the spatial context. All the factors of the existence of matter that also play roles in the existence of space must conform to their spatial equivalents. Beyond this, the immateriality of space allows for great latitude in other factors unique to matter such as substantiality, limited extension, motion, interaction, and changing pattern of organization, as long as they also conform to the nature of space.

Factors of the First Stage Required for the Existence of Matter

The manner in which these factors play roles in both space and matter is determined by the nature of what it is that exists in each case. The way they occur in the two cases is both the same and different. They must be the same because the factors of the one case must conform to their equivalents in the other case, and they are different because in each case what it is that exists is different.

For space to exist it must have continuous existential quantity, extension, and voluminality. Either that which exists has these factors and is-there, or it does not have these factors and is-not-there. For matter to exist it must have at least some continuous existential quantity, extension, and voluminality. The existential quantity of space is continuous immaterial place, while that of a primal unit of matter is continuous substance. The voluminal extension of space is that of immaterial place, while that of matter is the voluminal extension of substance.

There are eight other factors that play roles as qualities of continuous existential quantity, extension, and voluminality: parts, coexistence, relativity, sequentiality, distance, direction, organization, and place. The manner in which these factors play their roles in the case with matter conforms not only to their spatial equivalents, but also to those factors which are unique to the first stage of matter: limited, discontinuous, movable, interactive, substantial units with size, shape, surface, and topography. All of these factors interrelate in a manner dictated by the substantial mode-of-being existing within the spatial mode-of-being.

The continuous voluminal extension of spatial existential quantity is infinite. The continuous voluminal extension of material existential quantity is limited and discontinuous, occurring as individual units. In the case with matter, then, the eight common qualities play their roles within the basic or primal unit of matter, and the unique qualities of matter play their roles as aspects of limited discontinuous substantiality or as factors that occur because matter exists as units.

To exist, matter must have extension, and as with space, to have extension is to have parts, as for example one

Chapter 3: Space and Matter

half of the extension of a unit of matter and the other half. As with the parts of space, the parts of a unit of matter are coexistent, and thereby exist relative to one another. That relative coexistence has organizational factors. There is a sequential-difference from one part of the continuous material extension of the unit to another part. With this case of sequential-difference in a primal unit of matter there is a role for sequential enhancement. This role for sequential enhancement is an aspect of extensional development, an existential context dependent non-pathway factor development from the role of sequential enhancement that occurs with the sequential-difference of spatial extension.

Adjacent parts of the unit have sequentially based direction relations, but because they are adjacent they have no distance relations. With the material adjacent relation there is a contact relation between adjacent parts of the continuous substantial existential quantity of a primal unit. Adjacent relation plays a role with contact relations, and thereby with physical contact interaction, cause. Non-adjacent parts of the unit have both direction and distance relations, but because they have no direct contact relation there can be no direct interaction relations between them.

Similar to the case with immaterial locationally differentiated pattern, first level pattern, there is within a unit of matter a locationally differentiated pattern of material parts. This is a second, distinct, foundation of extensional pattern. The coexistent parts of a unit of primal matter have extensional relations between them. The relations are in addition to and existentially-dependent on the existence of the primary components of the coexistence, and that is combinatorial enhancement. Pattern is a combinatorial enhancement that involves the entire situation, the primary components and their extensional relations with one another. In this case it is three-dimensional pattern of direction and distance relations between parts of a primal unit, differentiated locationally and by the unique self-identity of the distinct parts of the unit. This is extensional pattern of coexistent parts of a primal unit of matter—sec-

Origins of Self-Organization, Emergence and Cause

Level 1   Level 2

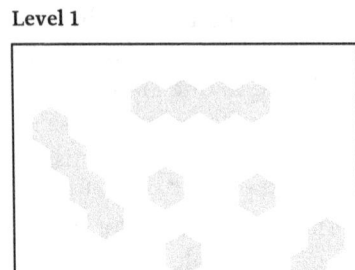 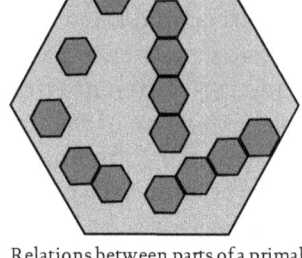

Relations between parts of space   Relations between parts of a primal unit of matter

Figure 3.1 *Material pattern of extensional relations intrinsic to a unit of matter—Second level pattern of extensional relations.*

ond level pattern (Figure 3.1). This stage of the development of pattern plays a role in the positional orientation relations between a static unit of matter and spatial place, third level pattern, and also in the positional orientation relations between the coexistent units of matter that materially differentiate fourth level pattern.

To have extension is to have magnitude. Space is extensionally infinite—matter is extensionally limited. While space exists as continuous, extensionally infinite, immaterial place, matter exists as internally continuous, extensionally limited substance. Because space is infinite it has no overall size, nor does it have a shape, outside edge or surface, or surface topography. An extensionally limited unit of matter has voluminal magnitude, a specific intrinsic size. A unit of matter, as a limited volume, also has an overall shape and a surface, and the surface, which is a form of extension, can have topography of some kind.

The voluminal extension of material existential quantity constitutes a form of place. Just as the extension of matter occupies and conforms to spatial extension, with that spatial extension thereby playing a role in the nature of the material extension, the form of place that is an aspect of material extension occupies and conforms to spatial place, with that spatial place thereby playing a role in the nature of the material place. The parts of material voluminal extension, the parts of a primal unit, constitute a form of place, a place for example, where a vibrational wave can pass through. This is place within a primal unit.

Just as voluminal extension has parts, whether of space or of matter, surface extension has parts, different areas of the surface. Surface area constitutes a form of place. As limited extension, size, shape, and surface are unique to matter, so also is this form of place which is associated with those factors. With the factor, surface place, goes the factor surface location. Surface place and surface location play important roles, especially in relation to topography. Surface configuration and location relations between surface features play roles in many more developed situations such as collisions between units of matter, molecular relations, weather patterns, and forensics.

Matter is movable. Immaterial space is static, there not being anything there to move. Spatial factors such as existential quantity, place, location, part, distance, direction, positional orientation, and organization are static, not moving, remaining always just as they are. With the substantial mode there is something there, and it moves. Virtually all matter is constantly moving. Those factors in material form are virtually always changing in relation to static space, in relation to their spatial equivalents.

Matter is interactive. It is not immaterial. An immaterial mode-of-being is so starkly simple that it does not have the wherewithal to interact with anything else that exists. An immaterial mode-of-being can interrelate with other modes-of-being in that it can be coexistent with them and thus exist relative to them in various ways, but it is intrinsically incapable of interaction. It can neither affect, nor be affected by, anything else. The substantial mode-of-being of matter, however, can be characterized in part by its capacity to interact one part with another. With the substantial mode-of-being something non-immaterial is-there, something not nearly so simple as immaterial space, something with a great capacity to affect and to be affected by. Factors such as the primal nature of matter, its existential quantity, size, shape, and surface features play major roles in material interaction.

Factors of the Second Stage (Continuing-Existence) Required for the Existence of Matter

Understanding the factors of spatial continuing-existence helps in understanding material continuing-existence, first because the continuing-existence of space shows in simple form what is foundationally required for there to be continuing-existence, its initiation and qualities, and second because spatial continuing-existence provides an existential context to which material continuing-existence must conform.

For space to exist there must be ongoing continuance of that existence. Either that which exists has continuing-existence and is-there, or it does not have this factor and is-not-there. For matter to exist there must be some continuance of that existence.

With matter, as with space, existence initiates existence, the existence of matter initiates the continuing-existence of matter. Space and matter are existential initiators, of which only three are known. Only space, matter, and motion initiate continuing-existence. The continuing-existence of all other factors, of all else that exists, is existentially based on the continuing-existence of the primal-forms-of-existence and on that of motion. For example, the continuing-existence of a pattern of material organization such as that of a population or an object is nothing more than the pattern of distribution in space of the continuing-existence of the matter that constitutes the existential basis of the population or object. Like space, matter is a primal-form existential initiator.

The factors that play roles in spatial continuing-existence play identical roles in the continuing-existence of matter. These factors are: initiation, consequent-existence, determinate-reality, uniform continuance, unidirectional change, newness, ever-increasing quantity without beginning, parts, sequentiality, non-coexistence, quantitatively specific after and before relations, relativity, organization and pattern, and development. There are two groups of factors here. There are those factors which play roles in the existence of matter, and which in developed form also play roles in its continuing-existence. There are additionally those factors that have their development-of-origin with the initiation of material continuing-existence. All together they constitute the developmental interrelation of factors that is material continuing-existence.

The continuing-existence of space and the continuing-existence of matter are identical except for two factors. First, immaterial place initiates spatial continuing-existence while the primal form of substance initiates material continuing-existence. Second, space is static, immobile, the continuing-existence of any particular part always occurring at the same location relative to all the rest of static space, while matter moves, the continuing-existence of any particular unit always occurring with the unit wherever that unit may be passing through.

The occurrence of material continuing-existence with spatial continuing-existence and the conformity of the material case with the spatial case play core roles in the deep structure of the process of emergence. These two distinct existential-pathway-developments are identical, and can have a one on one developmental interrelation of factors that plays a critically important role with emergence because the circumstances of each factor and its role are the same in both the spatial case and the material case. In Appendix 2 there is a summary of those factors that are identical in spatial and material continuing-existence.

Material continuing-existence is a development of change in an aspect of self-identity, and constitutes a separate foundational origin for this form of change. Just as in the case with space, during the change from one part of material continuing-existence to a following non-co-existent part, the continuance is maintained but the parts change. Because the parts of material continuing-existence have distinct self-identity, there is a continuous change in self-identity with the continuous change of the part of the continuance that is currently existing. Thus there is a continuous maintenance of the self-identity of that which is continuing to exist, matter, material existential quantity, while there is continuous change of the self-identity of the currently existing part of the continuing-existence of the matter.

The first form of the stages of change in an aspect of self-identity is, Change from one individually distinct part of continuing-existence to following non-coexistent individually distinct part. Of the three cases of this form, change in the self-identity of material continuing-existence is the second.

A factor can occur during one part of spatial continuing-existence and a different factor can occur during a different part of that continuance when the first factor no longer exists, the non-coexistent-sequential-difference of the continuance providing different separate parts of continuing-existence existential context for the non-coexistence of the two factors. Material continuing-existence, while it does not have a role as an existential context as in the case with space, it does have the same form of non-coexistent-sequential-difference, and provides for differences that occur in relation to matter, for example, in concert with spatial continuing-existence, changes of material organization.

The Identity of Matter Is Based on the Existence of Substantiality

Something is-there that is not immaterial. All the other factors of matter occur in relation to its substantiality. The substantial form of existence occurs as discontinuous units. The existential quantity, continuance, extension, and voluminality of the substantial mode-of-being are thus limited. The size, shape, surface features, and internal parts are factors of limited substantial units. The organizational aspects of these factors, such as the coexistent-sequential-difference from one side to the other and the internal distance and direction relations from one part to another, conform to the existence and nature of the substantiality—it is the sequential-difference of substance. The continuing-existence of matter, with its after and before organizational relations of non-coexistent-sequential-difference, is the continuing-existence of substantiality. And it is the substance of the unit that moves and interacts with the substance of other units.

These factors of the substantiality of matter are the factors of the existence of matter. For matter, to not have its substantiality and the qualities of that substantiality—to not have the existential quantity, continuance, extension, voluminality, continuing-existence, and intrinsic organizational relations of that substantiality—is to not exist.

Existence is the foundation of self-identity. Because the factors of matter exist, they are what exists, they are what they are. The existentially based, the existentially set, intrinsic self-identity of substantiality is the foundation of the determinate nature of the motion of matter and of material interaction.

Materially Based Extensional Development and Materially Based Change Development

Development is first encountered with the extensional development of space. It occurs again with the change development involved with spatial continuing-existence. With matter there is another foundational form of extensional development because, to exist, matter must have continuous voluminal extension. The extension of a unit of matter, and thereby material extensional development, conforms to the continuous voluminal extension of the space it occupies, and thereby conforms to spatial extensional development. Material extensional development differs from that of space in that it is based on substantial existential quantity, is confined to the material unit, is thus extensionally limited, and moves about through space with the unit.

The development from the existence of matter to the continuing-existence of matter is a development-of-origin for change development just as the equivalent transition is for space. For space it is an existential pathway, existentially-dependent, multifactor development that involves direct situation development and indirect factor developments. The situation with matter is nearly identical, differing in certain aspects due to the differences between the immaterial nature of space and the substance that is matter.

With space it is a consequent-existence existential-pathway-development of a single, extensionally continuous, infinite situation. With matter it is the consequent-existence existential-pathway-development of an apparently infinite number of extensionally limited discontinuous individual situations, the individual units of primal matter. Each unit of matter has a continuously ongoing individual intrinsic existential-pathway-development from its existence to its continuing-existence. All infinite space is coexistent and continues to exist simultaneously. Matter exists within the place-to-be provided by space, the extensional factors of matter existing simultaneously with the extensional space they occupy. Thus, all the apparently infinite units of matter are coexistent and all of their individual existential-pathway-developments occur simultaneously.

As with space, the situation with matter is an existentially-dependent development because it is consequent-existence, wherein the existence of the consequent is dependent on the existence of the initial factor. The continuing-existence of matter is existentially-dependent on the existence of matter. It is a direct development because the continuing-existence of matter is an immediate consequence of the existence of matter, without any intervening factors.

It is a situation development because change and enhancement occur. As with space it is change because it is the continuous transition from one part of material continuing-existence to another noncoexistent part that has distinct self-identity. And it is change in that the coexistent, three-dimensional extension of matter, by continuing to exist, gives rise to the noncoexistent, nondimensional, nonextensional factors of material continuance-of-being. It is change because factors occur with the development that are not constituents of the initial situation. It is a different development-of-origin for continuing-existence, consequent-existence, determinate-reality, change, newness, change development, and noncoexistence. The existence of these factors constitutes an enhancement of the situation.

There is indirect factor development because the factors of quantity, continuance, sequentiality, relativity, and organization play roles in both the initial situation and in the consequent, but just as in the case with space, the forms of the prior stage do not play roles in the forms of the developed stage. The modes in which the various factors occur in their first and second stages are distinct because the modes of existence and of continuing-existence are distinct. With matter, each stage of a factor derives its mode of existence either from the mode of substantial existence itself or from the mode of the continuing-existence of matter. The second stages do not derive their modes of existence from the modes of existence of the first stages, and as in the case with space, they are not direct developments of the first stages.

In this material existential-pathway-development, these indirect factor developments are existentially-dependent in that the developed forms of the factors are aspects of the continuing-existence of matter and are thus aspects of the continuing-existence of the initial forms.

Enhancement occurs with the indirect development of these factors in that with the consequent situation they each occur in a second, new, and different form. The consequent forms of these factors are developed forms because they are different from and existentially-dependent on the initial forms.

There are two distinct independent origins for change development, one each for the two primal-forms-of-existence. While they both are initiated by the existence of coexistent three-dimensional extension, those initial situations differ in that one is infinite immaterial place while the other is discontinuous extensionally limited substance. But in each case the consequent is nonexistent, nondimensional, nonextensional continuance-of-being. The qualities of material continuing-existence, its uniform, unidirectional, noncoexistent-sequential-difference with quantitatively specific after and before relations between the parts, are identical to those of spatial continuing-existence. Thus the change development that occurs in the two cases is essentially the same.

Space and Matter Are Coexistent—Matter Occupies Space

The situation of one unit of primal matter in space is, as far as is currently known, the foundation of unitized quantity.

A static unit of matter is coexistent with the space it occupies. Because space has parts, to occupy space is to occupy some part of space, some distinct individually unique spatial location. Matter has an occupation/location relation with the spatial place with which it is extensionally coexistent. Because matter exists as discontinuous units, the extent of the location/occupation relation matches the intrinsic extent of the particular unit.

The individually distinct parts of static space exist and thereby have existentially based intrinsic self-identity. Units of matter exist and thereby have their own existentially based intrinsic self-identity. The location/occupation relation between a static unit of matter and the spatial place it occupies is a relation between existentially based

Chapter 3: Space and Matter

intrinsic self-identities. The self-identity of the material unit occupies the self-identity of the spatial place. The unit of matter shares the placeness of the part of space it occupies. It has the same place as the part of space it occupies. A unit of matter occupying space is where it is, it has existentially based specific spatial location.

An extensionally limited static unit of matter is coexistent with all infinite space. That unit occupies an extensionally limited spatial place equivalent to the dimensions of the unit. That spatial place has specific quantitative direction and distance relations to all other spatial place. Thus, the material unit occupying that spatial place has the identical set of direction and distance relations to the rest of space. Because a unit of matter has an intrinsic shape, it has a positional orientation relation to space. The various sides of the unit face towards different directions in static space.

The spatial place occupied by a unit of matter has direction and distance relations with other spatial places, and thereby plays a role in immaterial locationally differentiated pattern—first level extensional pattern. The unit that occupies that spatial place has the same direction and distance relations with other spatial places, and thereby also plays a role in the existing pattern of extensional relations. But now that location is differentiated both locationally and by the matter. The pattern of extensional relations of the unit with the rest of spatial places is in part materially differentiated, at the location of the unit, and in part immaterially differentiated by location and unique self-identity of spatial place, at the other spatial places (Figure 3.2).

This is extensional pattern of coexistent primary components, space and one static unit of matter—third level extensional pattern, transitional level. It is static, three-dimensional pattern of direction, distance and positional orientation relations between a unit of matter and space. There is no factor of pattern in the coexistence relation between the unit and the space it occupies because in

Level 1

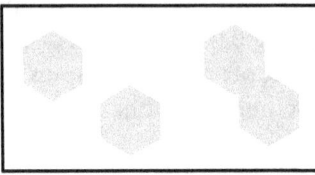

Extensional relations
between parts of space

Level 2

Extensional relations
between parts of a primal
unit of matter

Level 3

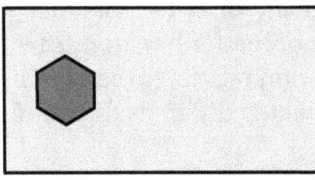

Extensional relations of a
unit of matter with space

Level 3

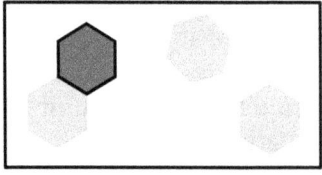

Extensional relations of a
unit of matter with
particular parts of space

Figure 3.2 *Third level extensional pattern, between a unit of matter and spatial place, differentiated in part by the substantiality of the unit and in part by the self-identity of the immaterial spatial places.*

the occupation relation of matter to space the location of the unit and the location of the space occupied are identical.

This is a transitional level in the development of extensional pattern. With first level pattern (between spatial places), all the primary components of the coexistence situation are immaterial and differentiated only by location and unique self-identity. With fourth level extensional pattern (between units of matter), all the primary components are units and are materially differentiated. Third level extensional pattern (between a unit of matter and spatial places) is a transitional level because the extensional relations are between a primary component that is differentiated materially and other primary components of the coexistence that are differentiated immaterially. Third level pattern is a developmental stage between a level where the extensional relations are between immaterial primary components differentiated only by location and self-identity, and a level where the extensional relations are between material primary components differentiated by self-identity and substantiality.

With coexistence there is relative existence and combinatorial enhancement. The occupation relation, and the direction, distance, positional orientation, and pattern relations of the unit to space are the enhancements of this coexistence.

Continuous voluminal spatial extension is infinite, constituting an infinite continuum of spatial place. Individual units of matter are extensionally limited. The continuity of individual material units is extensionally limited, discontinuous. Thus the distribution of matter is discontinuous, and matter does not form an infinite continuum. However, units or unitary systems of matter from photons to galaxies have been found in all regions of space that have been checked, and there is no reason to think that the distribution of the discontinuous units of matter does not match that of infinite space.

The extension of primal matter is coexistent with the extension of space. The material and the spatial existential quantities are coexistent. Substantial three-dimensionality is coexistent with spatial three-dimensionality. The continuance of material existential quantity is coexistent with the continuance of spatial existential quantity. The parts of a primal unit of matter are coexistent with the parts of space. And there is a coexistence of material coexistent-sequential-difference with spatial coexistent-sequential-difference.

The coexistence of space and a static unit of matter involves three stages of development for coexistence. There is the first stage, the coexistent parts of space. Then there is the second stage, the coexistent parts of matter.

And then there is the coexistence of these two cases, the first stage and the second stage existentially together. This third stage constitutes a merging, an interrelating, of two distinct existential-pathway-developments. Space and matter are not merely coexistent, they are interrelated by way of the existential context existential-dependency relation.

With the coexistence of space and matter goes the simultaneity of the continuing-existence of matter with the continuing-existence of space. Continuing-existence is a form of change. The noncoexistent-sequential-difference of continuing-existence provides for the possibility of other forms of change in relation to that which is continuing to exist. The continuing-existence of space provides for the possibility of change in association with spatial place, and the continuing-existence of matter provides for the possibility of change in association with matter.

With the coexistence of space and matter there is a simultaneous occurrence of the noncoexistent-sequential-difference of spatial continuing-existence with the noncoexistent-sequential-difference of material continuing-existence. This is a simultaneous occurrence of two cases of change. The simultaneity of the continuing-existence of space and the continuing-existence of matter provides for the possibility of change in some aspects of the relations between space and matter. This case of simultaneity is the dual continuing-existence existential context situation for other forms of change such as motion and changes based on motion, for example, emergence.

There are two distinct but coexistent primal-forms-of-existence, the continuing-existences of which constitute two distinct but simultaneous existential-pathway-developments. New part of material continuing-existence occurs simultaneously with new part of spatial continuing-existence. It is the new part of these simultaneous cases of change that provides the existential context foundation for the new part of other forms of change.

Development of Factors from Their Spatial Form to Their Material Form— Nonpathway Factor Development
Preliminary Considerations

Development itself develops. Infinite spatial place is the simplest, most primal, foundational factor of reality. The extensional development that is an aspect of spatial place is the foundational form of development. Spatial continuing-existence is the most primal foundational form of continuing-existence, providing a context for the existence of the other forms of continuing-existence, that of primal matter and that of motion. The occurrence of spatial continuing-existence as a consequence of the existence of space is the development-of-origin for change development.

There is a development from spatial extensional development to spatial change development. This is a case of existential pathway, existentially-dependent, direct situation, indirect factor development. The developmental factors (a) existential pathway, (b) existential-dependency, and (c) direct situation development naturally occur together. They are factors of this development because (1) spatial extensional development is an aspect of spatial place and (2) change development is an aspect of the transition from the existence of spatial place to its continuing-existence, and (3) because that transition is a case of consequent-existence. It is an indirect factor development because the mode of extensional development with no role for change is different from the mode of change development which is change. While change development is existentially-dependent on extensional development by way of direct situation development, it is only indirectly so as a factor development because the mode of the first stage does not play a role in the mode of the second stage.

There is another case of extensional development, that of the continuous extension of a primal unit of matter—material extensional development. Because the factors specific for the existence of immaterial place and those specific for the existence of substantiality are distinct and mutually independent of one another, this case of extensional development is foundationally different from that of space. Even though the foundations of the two cases are distinct, the extensional aspects of the existence of matter conform to the extensional aspects of the spatial place the matter occupies such that the factors of material extensional development do not differ greatly from those of spatial extensional development. There are some differences, for example the factors of the spatial case are static, unmoving and changeless, while the factors of the material case move about though space with the substance they are based on.

The existence of matter initiates the continuing-existence of matter, and a second case of change development occurs as an aspect of that transition. Like material extensional development, material change development is foundationally distinct from its spatial counterpart. Because all cases of continuing-existence are simultaneous and identical in their qualities, differing only in what it is that continues to exist, both cases of change development are essentially identical in nature, with the material case moving about with the material units. The developmental relations between material extensional and material change developments are the same as those for the spatial case.

To this point, then, there are two basic types of development, extensional and change, and there are two cases of each. From each of the two foundationally independent cases of extensional development there is an existential pathway case of change development. In both forms there is a role for continuance, the role of continuous extension in the one and the role of continuing-existence in the other. These roles of continuance are required for these forms of development.

Nonpathway Factor Development from Space to Matter

There are several factors that play roles both in space and in matter, with the material forms existing within and conforming to the simpler spatial forms. There are developments of these factors from spatial extension to spatial continuing-existence and from material extension to material continuing-existence, and in both cases there are factors of continuance involved. These factors develop also from the spatial form to the material form, but in this case there is no role for continuance. This is nonpathway factor development.

Nonpathway factor development is a different form of development from either extensional or change development in that there is no pathway connection between one stage and the next. With both extensional and change developments there are roles for continuous sequential-difference and existential pathway. With change development the roles of continuous sequential-difference and existential pathway involve a role for existential-dependency. With nonpathway factor development the stages are not connected by continuous sequential-difference, nor by existential pathway or existential-dependency. Another difference is that change development occurs by way of determinate consequent-existence, while nonpathway factor development has no role for these factors.

With change development a factor can occur both at a prior part of an existential pathway and again at a following part. In many cases it is an indirect development in that while the first occurrence of the factor in the pathway plays its role in what follows, the nature of the situation development is such that the first occurrence of the factor ceases to exist before the factor occurs a second time. The requisite set of factors that play roles in the development-of-origin of the factor in this particular pathway occur again at a following part of the pathway. Because their first occurrence is no longer present, the second occurrence is only indirectly dependent on the first occurrence, being so simply by way of the existential pathway of the situation development. The second occurrence of the set of factors that play roles in the existence of this factor is thus in large part independent of the their first occurrence. Yet there is an aspect of existential-dependency because it is the ongoing development of a single situation, a single existential-pathway-development.

Just as a set of factors that play roles in the existence of a factor can occur essentially independently more than once in the change development of a single situation, they can often occur completely independently in the change development of different situations. Nonpathway factor development is a development of a factor across from one developing situation to another, separate and distinct, developing situation.

It usually happens with change development that when a factor occurs at one part of a development of a situation and then again in a developed form at a following part of that situation development, the factor occurs in a developed form because there are a greater number of factors playing roles at that following part of the situation development. When a situation develops by way of an increasing number of factors playing roles, there are more factors available to play roles in the reoccurrence of individual factors and they thus tend to occur in developed form. Essentially the same thing can happen in separate unrelated situation developments. Because of the effects of such factors as (a) newly coming into existence, (b) coming together, (c) coherence, and (d) hierarchy, and the associated roles of sequential and combinatorial enhancements, a situation developing by way of existential pathway has a strong tendency to become progressively more complex. The form of factors in the earlier, less complex stages of the development tends to be simpler than in the later, more complex stages. Because a factor can occur independently in the developments of different unrelated situations, a factor can originate in a simple form in the early simple stages of one situation development, and originate again independently in an enhanced more developed form in a relatively later more complex stage of a different situation development.

This case of nonpathway factor development goes from space to matter because space is simpler and more fundamental than matter. Space is simpler than matter because immateriality is simpler than substantiality. An immaterial form of a factor is by way of its mode-of-being simpler in nature than is a material form of the factor. Thus there is a development here from simpler forms to less simple forms. Space is more fundamental than matter because space provides an existential context for matter. Matter does not provide an existential context for space. Space is existentially independent of matter, while matter is existentially-dependent on space. Space can exist without matter, while matter cannot exist without space. The context space provides consists of both extensional place and continuance-of-being. All the extensional factors of matter and all its continuance-of-being factors exist within and conform to their spatial counterparts. Because it is the factors of matter that conform to the factors of space, it is space that is the more fundamental form of existence.

The development from the spatial form of the factors to the material form is a development from one developmental pathway to another foundationally unrelated pathway. Space does not exist because of matter, and matter does not exist because of space. There are no factors of the existence of space which initiate factors of the existence of substantiality, nor are there factors of the existence of substantiality that initiate factors of the existence of space. Other than the existential context relation between them, the existence of space and the existence of matter are unrelated. Other than context, the extensional development of spatial extension is unrelated to the extensional development of material extension. Other than context, the change development of spatial continuing-existence is unrelated to the change development of material continuing-existence.

The development of factors from the spatial form to the material form is a nonexistential pathway, nonexistentially-dependent (except for existential context), noninitiated, nondeterminate, unconnected development. Unlike change development, nonpathway factor development is not significant in these ways. It is significant because it exists. Factors can become progressively more complex as they occur in progressively more complex situations, both within single existential-pathway-developments and across from one to another of various separate unrelated existential-pathway-developments. Nonpathway factor development is epistemologically important because of its utility for the overall understanding of a factor. When mapping an overview of the development of a factor, there are usually many instances of nonpathway development that fill out the complete sequence of stages. Furthermore, knowledge of the existence of this form of development alerts the mind to the possible occurrence of a factor at any stage of its development, whenever the appropriate circumstances for it occur, in any situation development, and at any stage or level of complexity of an appropriate situation development.

The development of extensional and continuing-existence factors from their spatial form to their material form is a nonpathway factor development with an aspect of existential-context-dependency. The factors that partake in this development are:

1. Extension;
2. Place;
3. Voluminality;
4. Continuance;
5. Quantity;
6. Parts;
7. Coexistence;
8. Relativity;
9. Coexistent-sequential-difference;
10. Sequential enhancement;

11. Existential-pathway-development;
12. Distance;
13. Direction;
14. Organization;
15. Combinatorial enhancement;
16. Initiation;
17. Consequent-existence;
18. Determinate-reality;
19. Existential-dependency;
20. Noncoexistence, and;
21. Noncoexistent-sequential-difference.

Space, matter, and motion—from these factors all else follows.

Chapter 4
Space, Matter, and Motion— The Origin of Emergence
The Existence of Motion

Motion is matter passing through space. There are no evident factors that necessitate the existence of motion. The existence of space requires neither matter nor motion. Nor does the existence of matter appear to require motion. Like matter, it is just there. Motion is something extra, a factor of reality that is in addition to space and matter. However, motion is existentially-dependent on matter, it is matter that moves, and the origin of motion is probably a factor associated with the unknown factors that necessitate the existence of matter.

The Intrinsic Nature of Motion
Factors of Self-Identity and Continuing-Existence of Motion

Either a unit of matter is at rest relative to space, or it is in motion relative to space. Either a unit of matter is not moving, or it is moving. With motion, as with anything else that exists, the difference between nonexistence and existence is complete. Because space is static, any motion relative to space is real motion. It is an existing aspect of reality, with existentially based intrinsic self-identity.

Any particular motion is the motion of a particular unit of matter through a particular part of space, during a particular part of the continuing-existence of the unit, and during a particular part of the continuing-existence of space. Motion occurs with the continuing-existence of the unit and the continuing-existence of space. Motion exists, and it continues to exist—it has continuance-of-being.

The continuing-existence of motion, like that of space and matter, is a continuous, uniform, unidirectional, sequential change, a form of noncoexistent-sequential-difference. As with the continuing-existence of both space and matter, newness occurs with the continuing-existence of motion, continuously new part of the continuing-existence of the motion, continuously new noncoexistent part. This is a third case of newness based on continuing-existence, and constitutes a developed stage in that both the continuing-existence of space and that of matter play roles in the continuing-existence of motion, spatial continuing-existence because it provides an existential context and material continuing-existence because it is matter that moves.

With the continuing-existence of motion there is another development of sequential enhancement as an aspect of change development. It occurs here in the form of new part of motion continuing-existence, and is existentially-dependent on both prior cases as motion continuing-existence is existentially-dependent on both prior cases of continuing-existence.

The continuing-existence of motion is another case of pattern based on noncoexistent-sequential-difference, the last of the three based on continuing-existence. Identical to the continuing-existence of space and to that of matter, motion continuing-existence has parts which have distinct self-identity differentiated by their individual existence and by their noncoexistence. The organizational aspects of that pattern are sequentiality, unidirectionality, and specific after and before relations between the parts. With both prior cases of continuing-existence playing roles in motion continuing-existence, this stage of pattern based on noncoexistent-sequential-difference is a developed form existentially-dependent on the prior forms.

As in the cases with space and matter, motion continuing-existence is a stage in the development of change in an aspect of self-identity. During the change from one part of motion continuing-existence to a following noncoexistent part, the continuance is maintained but the parts change. There is a continuous change in the self-identity of the part of the continuance-of-being of the motion that is current. There occurs a continuous maintenance of the self-identity of that which is continuing to exist, the motion, which occurs in relation to spatial extension, while there is continuous change of the self-identity of the currently existing part of its continuing-existence, which occurs in relation to spatial continuing-existence. The noncoexistent-sequential-difference of the continuing-existence of a motion, with its change of the self-identity of the current part, provides different separate parts of continuing-existence of the motion at which there can occur differences in relation to the motion, such as its occurrence at sequentially different parts of spatial extension.

Change in self-identity of motion continuing-existence is the third case of the first of the several forms of the stages of the development of change in an aspect of self-identity. It is the last case of the form, Change from one individually distinct part of continuing-existence to following noncoexistent individually distinct part.

Factors of Motion Itself— The Basic Intrinsic Qualities of Motion

Because the difference between rest and motion is complete, when a unit of matter is passing through space its motion is continuous. Like all forms of continuance, motion is a form of sequential-difference and has parts, like the part of the motion when the unit is passing through one part of space and the following part of the motion when the unit is passing through a different part of space. When a unit is moving through one part of space, the following part of the motion, when the unit will be moving through the adjacent part of space, does not then exist. And when the unit is moving through that adjacent part of space, the previous part of the motion, when the unit was moving through the previous part of space, no longer exists. The continuous sequentially occurring parts of motion are noncoexistent. Motion is a form of continuous noncoexistent-sequential-difference, a form of continuous change. Because matter passing through space is a form of continuous noncoexistent-sequential-difference with specific after and before relations between the parts, the motion itself is intrinsically unidirectional. The individually distinct parts of motion occur relative to one another with organizational qualities of sequentiality, unidirectional occurrence, and after and before relations.

Because the parts of an ongoing motion are noncoexistent, they have distinct individual self-identity. Because there is a continuous ongoing change of the current part of a motion, there is a continuous ongoing change of this aspect of the self-identity of the motion. This is the first stage in the development of change in an aspect of self-identity that goes beyond the stages of development of this factor that occur with the three cases of continuing-existence. It constitutes a second form of this factor: Change from one individually distinct part of motion to the following noncoexistent individually distinct part, of which there is only this one case. Because emergence is a change in the self-identity of a situation, this is an important stage in the developments toward the origin of emergence.

Motion provides two stages of the development of change in an aspect of self-identity, each of a different form. With space that which continues to exist is static—spatial place, spatial existential quantity, spatial extension, coexistent-sequential-difference. Change in self-identity occurs only as an aspect of spatial continuing-existence. Motion continuing-existence has a form of change in an aspect of self-identity that is essentially identical in its qualities to the change in an aspect of self-identity that occurs with the other two cases of continuing-existence, spatial and material. Like the three cases of continuing-existence, motion itself is a form of change, and as with that other form of noncoexistent-sequential-difference, there is both a maintenance of self-identity and a change of self-identity. While there is maintenance of ongoing motion there is ongoing change from part to noncoexistent part. There is a maintenance of the ongoing motion, of its self-identity as motion, and there is change of self-identity from one individually distinct part of the motion to another individually distinct part.

The stage of change in an aspect of self-identity that occurs with motion continuing-existence is existentially-dependent on the two prior stages, and has a one on one relation of ongoing change with them. Change in an aspect of self-identity of motion occurs in simultaneous one on one relation with change in an aspect of self-identity of motion continuing-existence. Thus, change in an aspect of self-identity of motion occurs in an existentially dependent one on one relation with the changes in an aspect of self-identity that occur with spatial and material continuing-existence.

As with the form of change based on the three cases of continuing-existence, there is a form of newness that occurs with the form of change that is motion, continuously new part of the ongoing motion. Because the three cases of change and newness based on continuing-existence play roles in the form of change and newness based on motion, and because motion itself is an additional factor in this development of change and newness, the change and newness of motion are developed forms.

Motion is a form of noncoexistent-sequential-difference with the continuous occurrence of new part, and as such is a development of sequential enhancement in the change development pathway. The noncoexistent-sequential-difference of motion and its case of sequential enhancement are existentially-dependent on the prior cases. The sequential enhancement of motion is the first stage of the development of this factor in the change development pathway the mode-of-being of which goes beyond that of the prior cases that were aspects of continuing-existence.

The three cases of change (noncoexistence-sequential-difference) that occur with the three cases of continuing-existence provide for the occurrence of difference or change of other factors in relation to that which is continuing to exist. Because motion is intrinsically a form of change, a form of noncoexistent-sequential-difference, motion itself is change in relation to other factors, (such as the coexistent-sequential-difference of space).

Because matter and motion are distinct factors of reality, and because motion is existentially-dependent on matter, when motion occurs, it does so in a coexistence relation with matter. This is an unusual coexistence relation because the nature of the relation is such that there are no secondary coexistence factors, and thus in this case no role for combinatorial enhancement intrinsic to the coexistence, all consequences of the relation being extrinsic to the relation. Other features of the coexistence are more standard. There is the simultaneous occurrence of the continuing-existence of motion with the continuing-existence of matter. There is, then, a simultaneity of new part of motion continuing-existence with new part of material continuing-existence. And there is the simultaneity of the noncoexistent-sequential-difference of motion continuing-existence with the noncoexistent-sequential-difference of material continuing-existence. Because new part of motion occurs simultaneously with new part of motion continuing-existence, in the coexistence of matter and motion, the newness and noncoexistent-sequential-difference aspects of the motion occur simultaneously with those aspects of material continuing-existence.

Because space and motion are distinct factors of reality, and because motion is existentially-dependent on space, when motion occurs, it does so in a coexistence relation with space. The coexistence relation of space and motion is different from the coexistence relation of matter and motion. Because it is matter that moves, any particular motion is always associated with a particular unit of matter. That is not the kind of relation motion has with space. Because of the coexistence of space and motion, motion has a location relation with spatial place, and because motion is a form of change, motion has a continuously new location relation with a sequence of different spatial place, a continuously new combining of the motion with different place. The location relation is a combinatorial enhancement, and because the combining is continuously new, there is continuously new enhancement of the situation.

The continuing-existence of motion occurs simultaneously with spatial continuing-existence. Thus new part of motion continuing-existence and the noncoexistent-sequential-difference of motion continuing-existence are simultaneous with their equivalent aspects of the continuing-existence of space.

Motion has intrinsic organization and constitutes a form of pattern. Motion is a pattern of noncoexistent-sequential-difference with parts and organizational relations between those parts that are isomorphic to the equivalent qualities of the other forms of pattern of noncoexistent-sequential-difference. Ongoing motion is existentially-dependent on the continuing-existences of space and matter, and the pattern of noncoexistent-sequential-difference of motion is thus existentially-dependent on this form of pattern as it occurs in those cases of continuing-existence, although it is not a direct existential-pathway-development from them. This is the fourth form of this kind of pattern, and these four forms constitute the foundation of all other forms of sequentially noncoexistent pattern. All more developed forms are relational in one way or another.

Space cannot influence motion. Since space is immaterial and cannot interact with matter, if there could be a situation in which an inert unit of matter that was not spinning was passing through a completely empty part of space, that unit would continue moving in a straight path at a constant speed. Motion that is not being influenced by another factor is uniform. Again, because space is immaterial, if there could be in a completely empty part of space an inert unit of matter that was not moving, that unit would remain without motion.

Motion Is an Initiator

Foundationally motion has two different roles as an initiator. One is that of an existential initiator like space and matter, while the other is that of a primal factor initiator. In both roles motion is the third factor that provides a foundational source of change. The existence of space and the existence of matter are foundational sources of change by way of their roles as existential initiators initiating the primal form of change that is continuing-existence. Space can have only this one role as an initiator, and thus only one role as a foundational source of change. Matter, at this stage of understanding, also has only this one role as an initiator, and only one role as a foundational source. But since there is still something awesome to be discovered about matter, further roles as a source of change should not come as a surprise. Like space and matter, motion is an initiator by way of its role as an existential initiator initiating continuing-existence. Motion however has an additional distinct type of role as a foundational source of change. It initiates ongoing motion.

Like space and matter, motion is an existential initiator in that it initiates continuing-existence. The existence of motion initiates motion continuing-existence. This also is determinate consequent-existence by way of continuing-existence of intrinsic self-identity, and here too the consequence is intrinsic to the initiator because the change is a continuance of self-identity. The continuance initiated by motion as an existential initiator involves intrinsic factors of motion continuance-of-being and occurs in relation to factors of material and spatial continuance-of-being.

Motion is a primal-factor existential initiator rather than a primal-form existential initiator. It is a developed stage of existential initiator, first because both prior existential initiators play roles in this stage, and second because this stage is a multifactor situation rather than the simple existence of a primal-form-of-existence. The existence of matter as an initiator plays a role because motion is existentially-dependent on matter, the initiation of the continuing-existence of motion thereby being existentially-dependent on the initiation of the continuing-existence of matter. The existence of space as an initiator plays a role because matter is existentially-dependent on space for a place-to-be, with the initiation of material continuing-existence existentially-dependent on the initiation of spatial continuing-existence for its role as an existential context. The initiation of motion continuing-existence is a developed stage of existential initiation because space, matter, and motion all play roles in the situation that is the passing of a unit of matter through space. The existence of motion is the third and the only other initiator, beyond the existence of space and the existence of matter, known to initiate a distinct case of continuing-existence.

Motion is a primal factor initiator in that it is a factor that occurs only in association with a primal-form-of-existence, matter. In the absence of matter, there can be no motion. In its role as a primal factor initiator, the existence of motion initiates ongoing motion.

The initiation of ongoing motion is determinate consequent-existence by way of continuing-existence of intrinsic self-identity. In the absence of disturbing factors, the motion maintains its intrinsic nature, it simply continues on. This case of continuance, that is initiated by motion as a primal factor initiator, involves intrinsic factors of the quantity of the motion, factors of the extent and the speed of the motion, and occurs in relation to factors of spatial existential quantity and spatial extension. Because the change here is a continuance of the self-identity of the initiator, the consequence is intrinsic to the initiator. Also because it is a continuance of self-identity, the initiation of ongoing motion is uniform. Undisturbed motion is uniform.

The initiation of motion continuing-existence is the third stage of initiation, and has the same form as the two prior cases. Again, it is consequent-existence by way of continuing-existence, wherein that which exists continues to exist as itself, with no roles for other factors, and the consequence is intrinsic to the initiator.

The initiation of ongoing motion is the fourth stage of initiation, and is the first change of form in that instead of change in relation to the context of spatial continuing-existence, it is change in relation to spatial extension. It is now consequent-existence by way of ongoing motion, wherein motion continues as itself, as the change that is motion, with no roles for other factors, and the consequence is intrinsic to the initiator.

Initiation Situation—Organizational Initiation

When a unit of matter exists in space, it has four basic types of extensional relations with spatial place—occupation/location, direction, distance, and positional orientation. These relations of the unit with spatial place are extrinsic to the unit. Motion initiates change of the extensional relations between a moving unit of matter and spatial place (Figure 4.1). Motion, however, is intrinsically existentially coexistent with the unit of matter that is moving. Because the extensional relations are extrinsic to the unit, changes in those relations are also extrinsic to the unit, and thereby to the motion. Because the motion is the initiator, the consequences in this case are extrinsic to the initiator.

When a unit exists in space, it is coexistent with spatial places. The unit and the spatial places are primary components of a coexistence situation. Where there is

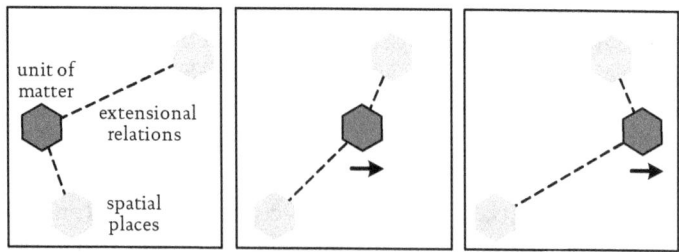

Figure 4.1 *Motion initiates change of the extensional relations between a unit of matter and spatial places. First stage of initiation situation—Organizational initiation.*

Chapter 4: Space, Matter, and Motion—The Origin of Emergence

coexistence there is relation, in this case extensional relations—which are organizational relations. The consequences of the initiator are not intrinsic to the initiator. They are changes in organizational relations between the unit and spatial place. Primary components of coexistence situations play roles in the nature of the relations between them. There are in this case of initiation three primary components playing roles in the nature of the consequences, the motion (establishing current location), and the unit and the spatial places (setting direction, distance, and positional orientation). The initiator provides the source of change within a context of factors extrinsic to the initiator. The initiator provides the source of change, but the nature of the consequences is in part determined by factors of the other two components, organizational factors.

When the initiator provides change within a context of extrinsic factors, there are consequences extrinsic to the initiator, and when the extrinsic factors and consequences involve extensional relations, relations of organization, then it is a case of organizational initiation. Organization plays a role in all stages of initiation. With the initiation of spatial continuing-existence there is the after and before relations between the sequentially occurring parts of that continuance. In cases involving the initiation of continuing-existence the organizational factors are consequences of the role of the initiator. Organization plays a greater role with the initiation that occurs with the motion of a unit of matter in relation to spatial place. There are organizational aspects of the relations between the primary components that play roles in the nature of the consequences of the initiation. Organization determines, in part, the nature of the consequences. This role of organization must occur if those particular consequence are to occur. With organizational initiation, the three initiators (space, matter, and motion) playing their roles together supply the source of change, but it is the extrinsic context organizational factors that determine the nature of that change.

Origins of Self-Organization, Emergence and Cause

When factors extrinsic to the initiator play roles in a case of initiation, and when there are consequences extrinsic to the initiator, when an initiator interrelates with other factors in the initiation of consequences, that is an initiation situation. The organizational initiation wherein motion initiates changes in the extensional relations between a unit and spatial place is the first stage of the development of initiation situation. In the development of initiation there are two major sequences of stages, (1) the stages of development of the three initiators (three stages of initiation of continuing-existence—of space, matter, and motion—and one stage of initiation of ongoing motion) wherein all consequences are intrinsic to the initiators, and (2) all stages thereafter, the development of initiation situations wherein there are extrinsic consequences.

Extrinsic to its occupation/location relation, a unit existing in space has three basic types of extensional relations with all space, with all spatial places—direction, distance, and positional orientation. Since spatial place is infinite and since the unit has extensional relations with every part of that unlimited extent, the unit has an unlimited number of each of these three types of extensional relations. All these relations together constitute the pattern of extensional relations of the unit with spatial place. As the unit moves through space, all these relations change simultaneously, continuously, from the currently existing pattern of extensional relations to a different subsequent pattern. Pattern of relations is a developed form of relation, and pattern of extensional relations is a fifth type of extensional or organizational relation of the unit to spatial place.

A unit of matter has a coexistence relation with infinite space, and an occupation relation with a specific extensionally limited spatial place. A unit has both a coexistence relation and an occupation relation with the place it occupies. Motion results in matter changing the spatial place it occupies. When matter moves, it is still coexistent with the place it left, with which it no longer has an occupation relation. Motion does not change the coexistence relation, but rather the occupation relation.

Since it is motion that supplies the origin of the change in initiation situation, the qualities of the consequences, which are also forms of change, are in many ways the same as the qualities of motion. The first of these qualities is consequent-existence. With the initiation situation, motion of a unit of matter in relation to spatial place, it is consequent-existence by way of the occurrence of the initiators in relation to extrinsic context factors. It is an organizational initiation situation, consequent-existence by way of the occurrence of the initiators in relation to extensional factors. It is consequent-existence by way of the occurrence of the motion of a unit of matter in relation to the occupation/location, direction, distance, positional orientation, and pattern of extensional relations of the unit to spatial place. During motion each different, consequent, extensional relation has unique self-identity distinct from the extensional relations that have gone before and from those that will follow. This constitutes a change in an aspect of the self-identity of the situation of a unit moving in space.

Continuing motion is determinate in that it is the ongoing existence of the intrinsic nature of motion. (What goes before, by way of continuing to be itself, determining what follows.) The changes in extensional relations are determinate in that they are consequences of the continuance of the self-identity of motion, and all its intrinsic qualities, in relation to the self-identity of space, and all its intrinsic qualities.

Because space is immaterial and thereby uniformly continuous, and because undisturbed motion is also uniformly continuous, the initiation and the occurrence of the changes in extensional relations occur in a uniformly continuous manner.

Just as there must be some quantity of the change of ongoing motion for there to be any motion at all, for there to be change of extensional relations between a moving unit and spatial place, there must be some quantity of change of extensional relations. If there is a continuous quantity, there are parts of that quantity. There are the changes in the extensional relations that are occurring when the unit is passing through one part of spatial place, and there are the different changes in those relations that are occurring when the unit is passing through a different part of spatial place. There is continuously new part of ongoing motion which initiates continuously new extensional relations of the moving unit to spatial place.

Combinatorial enhancement is the occurrence of relations between components of a coexistence situation. Prior development of combinatorial enhancement has involved already existing or static coexistence situations, coexistence relations that had no role for change. Motion is a form of change that initiates changes between a unit of matter and space. One of those changing relations is the occupation relation. Combinatorial enhancement occurs in the form of new occupation relation which occurs as a consequence of motion. The unit is newly combined with different spatial place. In this case occupation relation has a factor of togetherness that is in addition to that provided by mere coexistence, which can occur without the role of the occupation relation, as with locationally separate primary components.

Combinatorial enhancement occurs with any form of new combining by way of a factor of combination. The unit and the location it newly occupies were already coexistent before they were combined by way of motion. The enhancement in this case is not due solely to coexistence. The enhancement here is the new occupation relation, a newly occurring secondary coexistence factor due to the added factor, motion. The enhancement is initiated. With this stage of the development of combinatorial enhancement, where it occurs as a consequence of an initiator, it becomes a factor of change development.

Motion is a form of sequential-difference, and the relational changes it initiates are forms of sequential-difference. The sequentially occurring parts of motion are noncoexistent, making motion a form of noncoexistent-sequential-difference, and thereby the sequence of different extensional relations initiated by the motion also occurs as noncoexistent-sequential-difference. Noncoexistent-sequential-difference is change, and the change involved with these changes of occupation/location, direction, distance, positional orientation, and pattern of extensional relations is derived from the change of continuing-existence and ongoing motion. The change of continuing-existence and motion are initiated unidirectionally. Motion, whatever its direction through space, is always unidirectional in its manner of occurrence, and the changes of these relations are consequently likewise unidirectional.

The motion of a unit of matter in space changes the extensional relations of that unit to spatial place. There is a continuous change from one occupation/location relation to another, from one direction relation to another, from one distance to another, from positional orientation to positional orientation. Each newly occurring, sequentially distinct and noncoexistent extensional relation has individual unique self-identity, a different occupation/location relation, a different direction relation. The change from one extensional relation to a following noncoexistent extensional relation is a change in self-identity of the extensional relation situation between the moving unit and spatial place. The situation of each of these four types of extensional relation of the moving unit to spatial place is constantly undergoing a change in self-identity. This is a development of change of self-identity from the case of motion with its change from one part of the motion to a following noncoexistent part.

When motion changes the extensional relations of a unit of matter with spatial place, it thereby changes the pattern of those relations. A change of the extensional relations between the components of a pattern, in this case a materially differentiated unit and all the different locationally differentiated spatial places, constitutes a change from one pattern to a following noncoexistent pattern. This constitutes a change in the self-identity of the situation of pattern of relations between the unit and spatial place. Again this is a development of change of self-identity, of a situation rather than a single factor like motion, and all prior cases of change of self-identity play roles in this stage. Both of these stages of change of self-identity play roles in the next stage, the change of self-identity that occurs with emergence, which is also a change in self-identity of a situation.

Sequential enhancement occurs with changing extensional relations between a moving unit and spatial place. A new occupation/location relation, for example, or a new distance relation, is an enhancement of the situation—an aspect of the situation that was not there before. The noncoexistent-sequential-differences in extensional relations of a moving unit of matter to spatial place, and the noncoexistent-sequential-difference in the pattern of those relations, are stages in the development of sequential enhancement in the change development pathway. Because these changes are occurring in relation to spatial place, and because they involve progressively more spatial extension, and thereby extensional development, there occurs a one on one developmental interrelation between the extensional and the change developments of sequential enhancement. This situation is existentially-dependent on the roles of the prior cases of both developmental pathways.

The series of continuously new extensional relations between a moving unit and spatial places occur relative to one another. They do so with sequential after and before relations. The changes of occupation/location, distance, direction, and positional orientation relations between a moving unit and space occur in a pattern of noncoexistent-sequential-difference, as does the continuous change of pattern of extensional relations. There are, here, two developments of the pattern of change (change in extensional relations and change in pattern of those relations), and the four prior stages (three of continuing-existence and one of ongoing motion) all play their roles here. Then these two stages play roles in further developments of pattern of change, for example, the development-of-origin of emergence. The patterns of noncoexistent-sequential-difference, with after and before relations, of the sequence of extensional relations between a moving unit and spatial place are a form of organization.

The various types of changes of extensional relations initiated by motion in the context of spatial place play roles in relation to other factors in accordance with the natures of their intrinsic qualities of change.

Motion Is a Developed Form of Change

Motion is a developed form of change because it is compound change, because with motion there are two fundamental sources of change, continuing-existence and motion itself. It is a developed form of change because it has a greater number of factors and relations than occur with any case of the simple change of continuing-existence. The sixteen most significant factors that play roles in the existence of motion play more than sixty-four roles of thirty-five different forms. These factors play essential roles:

1. Extension (one role, one form);
2. Continuance (five roles, three forms);
3. Uniformity (five roles, three forms);
4. Parts (five roles, three forms);
5. Static and unchanging (one role, one form);
6. Distinct individual self-identity (five roles, three forms);
7. Coexistence (four roles, three forms);
8. Sequentiality (five roles, three forms);
9. Initiators (four roles, two forms);
10. Continuing-existence (three roles, one form);
11. Noncoexistence (four roles, two forms);
12. Change (four roles, two forms);
13. Change of self-identity (four roles, two forms);

14. Unidirectionality of change (four roles, two forms);
15. Newness (four roles, two forms), and;
16. Simultaneity (six roles, two forms).

Motion is a developed form of change because of development, isomorphism, and interrelating existential-pathway-developments. Many of the sixteen factors develop; there are isomorphisms between various factors and between various combinations of factors; the sixteen factors, sixty-four roles, and thirty-five forms of those roles all occur together in the motion situation; and all the factors, roles, and forms interrelate with one another. In Appendix 3 there are two lists of the factors that play roles with motion, a list organized by individual factor, and a developmental list of the factors and their roles based on the developments from space to matter to motion.

The Role of Isomorphism

Motion is required for emergence because it is motion that initiates change in the occupation/location relation of a unit of matter to space, and thereby change in the direction, distance, and positional orientation relations of that unit to the rest of spatial place (the places that can be occupied by other units). During the noncoexistent-sequential-difference of the simultaneous continuing-existence of motion, matter, and space, the noncoexistent-sequential-difference of motion sequentially relates a unit of matter in a foundationally unidirectional manner to the coexistent-sequential-difference of spatial extension. The noncoexistent sequentiality of continuing-existence provides the factor, noncoexistence, for developed forms of change such as motion, relational changes, and changes in the self-identity of a situation. It is the role of isomorphism between the roles of sequential-difference of continuing-existence, motion, and spatial extension that makes it possible for motion to occur and thereby for the consequent changes in extensional relations between a unit of matter and space to occur.

Chapter 4: Space, Matter, and Motion—The Origin of Emergence

Spatial extension is a uniformly continuous, nonunitized, sequence of individually unique coexistent part—a case of coexistent-sequential-difference. Continuing-existence, in all three cases (space, matter, and motion), is a uniformly continuous, nonunitized, sequence of individually unique noncoexistent part—a case of noncoexistent-sequential-difference. Undisturbed motion, also, is a uniformly continuous, nonunitized, sequence of individually unique noncoexistent part—again a case of noncoexistent-sequential-difference. There is an isomorphism between the qualities of these five cases of continuous sequential-difference (Figure 4.2).

The three cases of continuing-existence are simultaneous and isomorphic, with a one on one existential relation between parts. New part of motion is simultaneous to new part of motion continuing-existence and the qualities of motion are isomorphic to the qualities of motion continuing-existence, with a one on one existential relation between the parts of motion and the parts of the continuing-existence of motion. Because motion continuing-existence has an isomorphic one on one relation both with motion and with the continuing-existences of space and matter, there occurs an isomorphic one on one relation between motion and the continuing-existences of space and matter, with new part of motion simultaneous to new part of material continuing-existence and new part of spatial continuing-existence. There are together here four cases of noncoexistent-sequential-difference, four cases of continuous, uniform, unidirectional change, with sequential one on one relations between all of their parts.

Spatial extension is a case of continuous uniform sequential-difference. Motion occurs in relation to space. The continuous uniform sequential-difference of motion occurs in isomorphic relation to the continuous uniform sequential-difference of spatial extension. The noncoexistent-sequential-difference of motion occurs in relation to the coexistent-sequential-difference of spatial extension. Sequentially new part of motion is continuously occurring

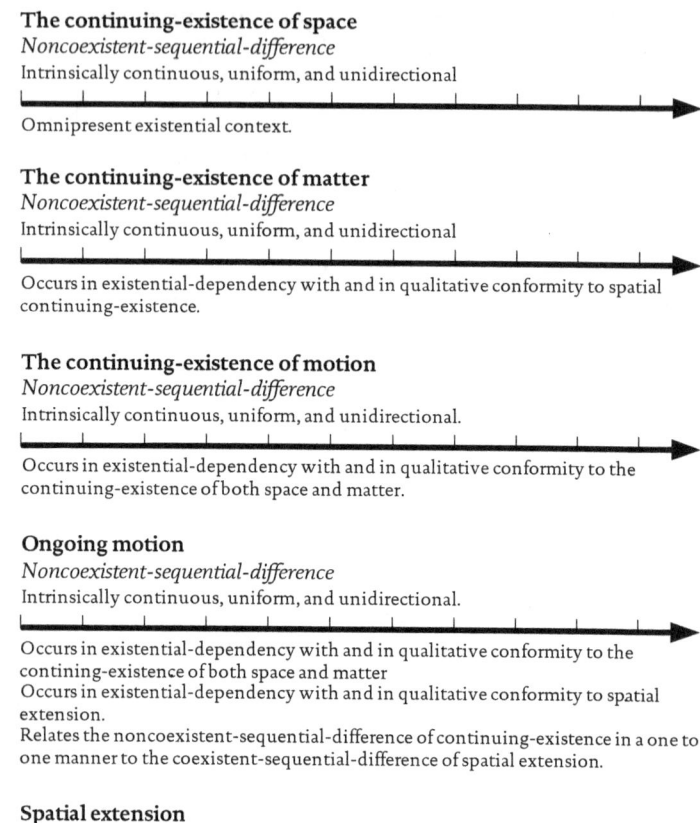

Figure 4.2 *The isomorphism between five cases of continuous sequential-difference makes it possible for there to be developed forms of change such as emergence, situation development, and cause.*

in relation to sequentially different part of space. There occurs a one on one relation between the changing noncoexistent parts of motion with the static coexistent parts of space.

Motion itself occurs in the sequentially coexistent existential context of spatial extension, and motion continuing-existence occurs with the sequentially noncoexistent existential context of spatial continuing-existence. Because of the isomorphism of these different modes-of-being, the noncoexistent sequentiality of motion occurs in one on one relation both with the noncoexistent sequentiality of continuing-existence and the coexistent sequentiality of spatial extension. Motion thereby sequentially relates a unit of matter to the sequentiality of spatial extension during the sequentiality of continuing-existence. With motion the two developmental pathways of pattern, extensional pattern and pattern based on noncoexistent-sequential-difference, developmentally interrelate.

The Role of Continuing-Existence in Developed Change

All forms of change are forms of noncoexistent-sequential-difference, that is, they have sequentially occurring individually distinct noncoexistent parts. With the case of spatial continuing-existence, there is a maintenance of unchanging self-identity during the change of continuing-existence. Because continuing-existence is a form of noncoexistent-sequential-difference, there can occur differences in relation to the unchanging self-identity at different parts of the continuing-existence of that unchanging self-identity. For example, a spatial location can be occupied by a unit of matter at one part of the continuing-existence of the spatial location, but not occupied by that unit at a different part of the continuing-existence of the spatial location. Differences in relations occur at different parts of continuing-existence. The noncoexistent-sequential-difference of continuing-existence is required for there to be relational differences to a spatial location. Differences in relations can occur only at different parts of continuing-existence.

With motion there is a maintenance of self-identity, a maintenance of motion itself, during the noncoexistent change from part to part of the continuing-existence of the motion. Because the continuing-existence of motion is a form of noncoexistent-sequential-difference, there can occur differences in relation to the motion at different parts of the continuing-existence of the motion. For example, a motion occurs in relation to one spatial place at one part of the continuing-existence of the motion and occurs in relation to a different spatial place at a different part of the motion's continuing-existence. As in the case with space, differences in relations between motion and spatial places can occur only at different parts of continuing-existence.

Motion is itself a form of noncoexistent-sequential-difference. With motion there is both a maintenance of self-identity and a change of self-identity. There is a maintenance of its self-identity as motion, and there is continuous change of self-identity from one part of the motion to another individually distinct part.

Motion can have differences in an aspect of its self-identity as motion because motion continuing-existence is a form of noncoexistent-sequential-difference, and is simultaneous and isomorphic with spatial continuing-existence. Distinct part of motion occurs at distinct part of the continuing-existence of the motion and simultaneously at distinct part of the continuing-existence of space. Differences or change in relation to space or to matter can occur only at different parts of the continuing-existence of space or of matter. At a stage of development of change in an aspect of self-identity, where it is a change in an existential factor rather than a factor of continuing-existence, that change can occur only at different parts of the continuing-existence of the existential factor.

The continuing-existence of space, matter, and motion provides the noncoexistent-sequential-difference required for the existence of other forms of change such as motion with its change in an existential factor, relational change, and change in an aspect of self-identity of a group or pattern. Change in an aspect of the self-identity of a group or pattern is existentially-dependent on relational change, which is existentially-dependent on motion. Motion is dependent on motion continuing-existence, which is dependent on material continuing-existence, which is foundationally dependent on spatial continuing-existence. With the stages of the development of change in an aspect of self-identity there is an accumulative playing of roles by prior stages in later stages, with everything dependent on the roles of continuing-existence.

The process of emergence is a process of change of self-identity, and the process of emergence, change in pattern of organization, is existentially-dependent on motion. Motion, in its role as an initiator, is the foundational source of the change in self-identity that is more developed than that which occurs with continuing-existence. Motion is thus, in an emergent situation, the foundational source of change in the self-identity of factors other than continuing-existence.

Space and Two Static Units of Matter

The situation of two coexistent units is a development of unitized quantity. The development of unitized quantity occurs by way of sequential enhancement, more or additional units, more or additional unitized quantity. This can be sequential enhancement as in extensional development, for instance the sequence of beads along a string, or it can be change development, in association with extensional enhancement, as with threading additional beads onto the string.

The coexistence of two static units of matter in space constitutes a development of coexistence as a factor of reality, and thus a development of combinatorial enhancement. With coexistence comes relative existence, and with relative existence comes enhancement in the form of relations between primary components of the coexistence. This is combinatorial enhancement. One factor of enhancement that goes with this stage of the development of quantity is the development-of-origin for the factor group. Proximity is an aspect of this factor. Groups can be quite dispersed, but the closer the units are the more group-like the situation is.

Two units coexistent in space have extensional relations between them of distance, direction, and positional orientation. These relations together constitute the pattern of extensional relations between the two static units. This extensional pattern is materially differentiated by the extensionally limited units. The direction, distance, positional orientation, and pattern relations between the units are the enhancements of this coexistence (Figure 4.3).

The units can be adjacent to one another in space, or nonadjacent. The combinatorial enhancement that goes with each of these stages is different. With adjacent relations the materiality of the units has the effect that the units are in contact with one another. Contact relations make it possible for there to be interaction between the units, for there to be causal relations between the units. The development-of-origin of cause is described in Chapter 5. In addition to the contact relation, adjacent units have a directional relation between them, but because there is no space between them when they are adjacent, there is no distance relation. There is also a positional orientation relation between these adjacent units which involves

Level 1

Relations between nonadjacent and adjacent parts of space

Level 2

Relations between parts of a primal unit of matter

Level 3

Relations of units with a nonadjacent spatial place and with an adjacent spatial place

Level 4

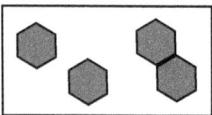

Relations between nonadjacent and adjacent units of matter

Figure 4.3 *Levels of extensional relations.*

which side of one unit is facing which side of the other and which part of one unit is nearer or farther from the parts of the other. Two static adjacent units of matter with their directional and positional orientation relations constitutes extensional pattern of adjacent primary components. The intrinsic extensional factors of the one unit are coexistent with the extensional factors intrinsic to the other unit. The voluminality, parts, and organizational factors such as coexistent-sequential-difference, direction, and distance relations of the parts of the one unit are coexistent with these factors of the other unit. Thus there are direction relations between the parts of the two separate units. And there are distance relations between the parts of the two units that are not playing roles in the direct contact relation. These relational factors of extensional pattern between the parts of the units play roles at later stages of development when the units move and interact.

With nonadjacent relations the units are not in contact with each other, and there cannot be any direct causal interaction. There are direction relations between the units, and because they are not adjacent, not in contact, there are distance relations. There is the positional orientation relation between nonadjacent units, and direction and distance relations between the parts of the one and the parts of the other. And finally, two static nonadjacent units with their direction, distance, and positional orientation relations constitutes extensional pattern of nonadjacent primary components.

When playing in the snow and throwing snowballs, it is not necessary to waste time and energy dodging a snowball that is not going to hit you. Adjacent and nonadjacent relations, and the roles they play with contact and interaction or noncontact and noninteraction, are of particular importance in the structure and function of reality at later stages and higher levels of development.

There is an existential-dependence relation between the pattern of relations of the units and the pattern of relations of the spatial places the units occupy. When

Origins of Self-Organization, Emergence and Cause

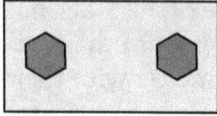
Units of matter with extensional relations with spatial places

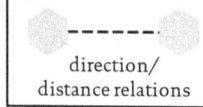
Two particular spatial places

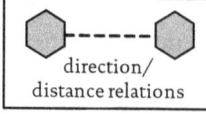
Two units of matter at two particular spatial places

Figure 4.4 *Existential-dependence relation of material pattern of relations on spatial pattern of relations.*

two units of matter are coexistent in space they each individually have all the relations with space that were described previously, but specifically they each have an occupation/location relation with the spatial place they each occupy, and direction, distance, and positional orientation relations with the rest of spatial place. Because they each occupy a specific spatial place, and because these two spatial places have specific direction and distance relations between them, the two units of matter have between them exactly the same extensional relations. The extensional factors of the immaterial spatial level of pattern are also the extensional factors of the material level. The direction and distance factors of the level of spatial pattern are also the direction and distance factors of the level of material pattern. This because anything that exists, like two units of matter, must exist in space, occupy spatial place, and share with the occupied place all of its extensional relations with other spatial place (Figure 4.4).

The coexistence of two units of matter in space constitutes a development of extensional pattern as a factor of reality. The immaterial locationally differentiated pattern of the occupied spatial places is at this stage materially differentiated at both locations. The mode-of-being of extensional pattern at this stage has two distinct levels of existence, two levels of organization. There is the level of spatial place with immaterial pattern differentiated locationally and by unique self-identity—first level extensional pattern. This is the level that provides the foundational factors of extensional pattern. Then there is this level of matter with pattern materially differentiated by two static units—fourth level extensional pattern.

The situation of two coexistent units constitutes fourth level extensional pattern, and has the following two stages. With both stages it is material extensional pattern differentiated by extensionally limited units. At the adjacent stage, there are roles for direction and positional orientation, which in turn play roles in the patterns of the contact and interaction relations. At the nonadjacent stage, the pattern of relations has aspects of direction, distance, and positional orientation, but the role of the distance relation prevents the occurrence of contact and the roles of direction and positional orientation therein.

Because the units are coexistent in the existential context of space, there is the simultaneous occurrence of the continuing-existence of the one unit with the continuing-existence of the other unit. There is a simultaneity of the noncoexistent-sequential-difference of the continuing-existence of the two units, with new part of the continuing-existence of the one simultaneous to new part of the continuing-existence of the other. The simultaneity of the continuing-existence of the one unit with the continuing-existence of the other unit provides for the possibility of change in some aspects of the relations between the units.

Motion of a Unit of Matter in the Presence of Another Unit That is Not Moving

Either a unit of matter is at rest relative to space, and thereby to everything that exists, or it is in motion relative to space, and thereby to everything else that exists. The motion of matter through space has the consequence that spatial relations between units of matter change (except for the case of isomorphic motion, the motion relation between units that have equal speed in the same direction). When the spatial relations between units of matter change, the patterns of organization of those units change. Motion reorganizes matter. Patterns of organization come newly into existence—they emerge. The coming into existence of a pattern of material organization due to the occurrence of motion and the consequent change in spatial relations between units of matter is the development-of-origin of emergence.

Factors of This Stage

The motion of a unit in relation to a stationary unit is an initiation situation, and at this stage it is again a case of organizational initiation. It is a coexistence situation between three primary components, space, a static unit of matter, and a moving unit, with all the organizational factors that involves. All the factors and relations that play roles in the situation of a static unit in space are here. All the factors and relations that play roles in the situation of a unit of matter moving in space are here. And all the factors and relations that play roles in the situation of two coexistent units that are not moving are here. What is different in this situation is that the motion of the one unit occurs not just in relation to the rest of spatial place, including the particular spatial place occupied by the static unit, but the motion occurs also in relation to that static unit. It is organizational initiation, a developed stage, because the initiator, motion, occurs in relation to extrinsic factors, a second unit of matter and the organizational extensional relations between that unit and the moving unit. The consequences are changes in organizational factors, changes in extensional relations. As in the prior case of organizational initiation, at the third level, changes in extensional relations here are extrinsic to the moving unit, and thereby extrinsic to the initiator, motion.

This is consequent-existence by way of motion of a unit of matter relative to another unit that is not moving. The consequents are new distance, direction, and positional orientation relations between the units. The two units and the extensional relations between them constitute an extensional pattern. When the extensional

Chapter 4: Space, Matter, and Motion—The Origin of Emergence

relations change, the pattern changes. New extensional pattern is the emergent consequence of relative motion between units of matter.

Like all cases of consequent-existence, this one is determinate. The initiators are cases of determinate change—with determinate intrinsic consequences—because they are cases of continuing or ongoing existence—the continuing being of space, matter, and motion, the continuance of intrinsic self-identity. With the first case of initiation situation, organizational initiation of changing extensional relations between a moving unit and spatial place, the moving unit contributes an aspect of determinate change, while the spatial place contributes to the situation an aspect of static, unchanging, self-identity. The determinate change occurs in relation to an extrinsic factor of existentially based self-identity, and therefore with determinately initiated consequences extrinsic to the initiator motion.

The case of a unit moving in relation to a stationary unit is based on that prior stage, with the same initiator contributing an aspect of determinate change. In this case, it is the stationary unit that contributes an aspect of static self-identity. The determinate change occurs in relation to this additional extrinsic factor. Because of the occupation/location relation between the stationary unit and the occupied spatial place, the extrinsic determinately initiated consequences are organizationally the same as in the prior case.

As with the prior case of organizational initiation, the consequences are forms of change. And like that prior case, these changes have a certain set of qualities, the same set as the prior case because of the existential-dependency relation between the two stages, with differences only as those changes are now in relation to a static unit occupying a specific spatial place. For emergence, that difference is all important, for these are the intrinsic qualities of change that play roles in the development-of-origin of emergence.

There is a continuous uniform initiation of new part of ongoing change such that there occurs an increasing quantity of change of extensional relations between the units. The parts of changing extensional relations are initiated sequentially and are noncoexistent, such that the change of these relations is unidirectional noncoexistent-sequential-difference. Since the parts of continuously changing extensional relations between the two units are noncoexistent and have individual unique self-identity, the extensional relations are continuously undergoing a change of self-identity. New part of an ongoing change with unique self-identity is a form of sequential enhancement, and the changes of extensional relations between the moving unit and the stationary unit are change development, the next stage from that of changing extensional relations of a moving unit and spatial place. Change in extensional relations between the units is a change in an aspect of self-identity of the situation. At this stage of development it is a change in an aspect of self-identity of the group of two units. This stage of change development is emergence.

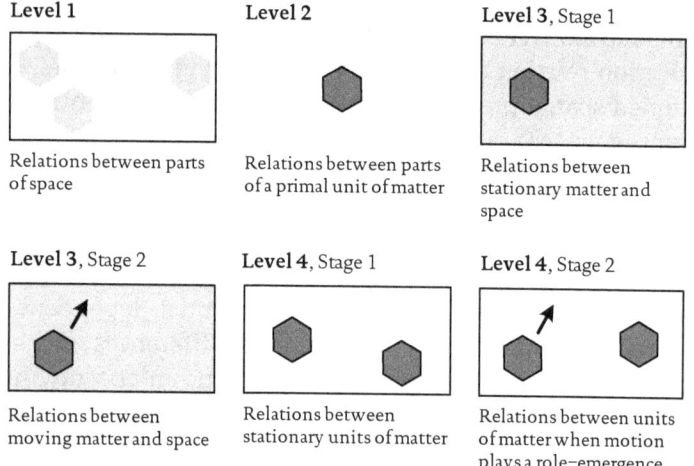

Figure 4.5 *Six stages of the development of extensional pattern and the emergence of new pattern of material organization.*

As in the prior stage of initiation situation, there are here two stages of development of change development. There is the stage of change development that occurs with the change of an individual extensional relation, such as a change of direction relation, where the occurrence of new direction relation between the units is an enhancement of the situation. And there is the stage of change development that occurs with the change of pattern of extensional relations, where the occurrence of a new pattern of extensional relations between the units is an enhancement.

The noncoexistent parts of ongoing change of extensional relations at this the fourth level of extensional pattern occur sequentially relative to one another, and have after and before relations that constitute a pattern of noncoexistent-sequential-difference. The ongoing change of extensional relations has intrinsic organization.

Discussion of Factors at the Origin of Emergence

The process of emergence, at its development-of-origin, is a change in extensional relations between units of matter, a change in the pattern of the extensional relations. There are six stages of the development of extensional pattern here that play significant roles, and four levels of that development. Two of the levels have two stages of development, a stage with static components where change plays no role, and a stage with a moving unit where change does play a role (Figure 4.5).

1. There is the stage of spatial place, with immaterial pattern differentiated locationally and by unique self-identity, providing the existential context level for the developed stages—the first level.

2. There is the stage of a primal unit of matter, with material pattern differentiated again locationally and by unique self-identity, which plays a role in the positional orientation of a unit to spatial place and to other units—the second level.

3. There is the stage of a static unit in space, with a pattern of extensional relations that is in part materially differentiated and in part immaterially differentiated by location, a transitional stage wherein the first level begins to provide extrinsic extensional relations for fourth level primary components—the third level.

4. There is the stage of changing pattern of relations of a moving unit to space, a second stage of the third level, a stage where motion continuously initiates change in the extensional relations of the unit to spatial place, thereby continuously changing the extensional relations between the materially differentiated part of the pattern to the immaterial parts differentiated by their static location.

   (Here the two distinct existential-pathway-developments of pattern combine, with the development of pattern based on noncoexistent-sequential-difference interrelating with the development of extensional pattern.)

5. Then there is the stage of the development of pattern that consists of two static units in space, with material pattern differentiated by distinct, extensionally limited units—the fourth level.

6. And finally, there is the stage of changing pattern of relations of a moving unit to a static unit, a second stage of the fourth level, a stage where motion initiates change in the extensional relations between fourth level components.

   (Again, because the continuous change of extensional relations between the moving unit and the static unit is a development of pattern of noncoexistent-sequential-difference, and because this also is an interrelating of the two distinct pathways of pattern development [extensional and change], this is a development, a second stage, of that developing interrelated pathway.)

Extensional development can occur by way of change development. Extensional pattern can develop by way of change development, as in extensional development by way of additional units. In like manner an extensional pattern can develop by way of the change that is reorganization. The first stage of extensional development by way of change development is the motion of a unit in relation to spatial place. Extensional pattern develops here by way of newly occurring extensional relations, by way of newly occurring pattern of extensional relations, and by way of the change in self-identity of the moving unit's pattern of extensional relations with spatial place.

As was discussed previously, in the sections on the relation of a moving unit to spatial place, in the development of extensional pattern, where it first interrelates with pattern of noncoexistent-sequential-difference and change occurs in extensional relations between the primary components of the pattern, between a moving unit of matter and spatial place, the change is initiated by the moving unit, not by the static immaterial places. In the next stage of this interrelated development, there occurs change in extensional relations between two primary components both of which are of the fourth level of pattern development. Again in this case the change in extensional relations is initiated by the moving unit, and not by the static unit. The static unit, because it is not moving, is incapable of initiating change in extensional relations—it lacks the wherewithal to do so.

The change in extensional relations between the units is initiated only by the moving unit, but the change is relational in nature and occurs between the primary components of a coexistence situation, between the components of a fourth level pattern. It is a change in the organization of a fourth level pattern situation. The factor, motion, that initiates the change does so in the context of fourth level pattern organization. It is an initiation situation, an organizational initiation, wherein the consequences are the result of motion initiating change at fourth level pattern which is fully differentiated by units of matter. The factor of motion alone, as in the third level pattern situation, is not sufficient to effect organizational change except in the context of extrinsic factors with organizational relations between them. With this stage of organizational initiation, it is that added unit and the organizational factors that go with it that gives this stage its character. In this case, as in the prior case, the initiators play their roles, but it is the organizational factor of fourth level pattern that makes possible the form of change that occurs. With organizational initiation, the initiators initiate the change, but the organizational factors determine the nature of the change.

Because the two units of matter are coexistent, the continuing-existence of the one is simultaneous to the continuing-existence of the other. New part of the continuing-existence of one unit is simultaneous to new part of the continuing-existence of the other. With the previous case of a moving unit of matter alone in space, there were three isomorphic cases of simultaneous continuing-existence, that of space, that of matter, and that of the motion, with simultaneous occurrence of new part of each of them. The noncoexistent aspect of spatial continuing-existence made it possible for there to be differences in relation to space at different parts of the continuing-existence of space. The noncoexistent aspect of material continuing-existence made it possible for there to be differences in relation to the unit at different parts of the continuing-existence of the unit. And the noncoexistent aspect of motion continuing-existence, plus the noncoexistent aspect of motion itself, together made it possible for there to be differences in relation to motion at different parts of the continuing-existence of motion. The simultaneity of the noncoexistent aspect of these four cases of isomorphic noncoexistent-sequential-difference made it possible for there to be differences in the extensional relations between the moving unit and space. With the case of the fourth level pattern of two coexistent units, one of which is moving, all of these prior factors still play their roles, but there are now the roles of the added factor, the second, static, unit. The simultaneity of the noncoexistent aspect of the continuing-existence of the static unit with the noncoexistent aspect of the prior four cases of noncoexistent-sequential-difference makes it possible for there to be differences in the extensional relations of direction, distance, and positional orientation between the two units of matter.

With fourth level pattern of extensional organization, the simultaneity of the noncoexistent aspect of the continuing-existence of the components of a pattern with the noncoexistent aspect of motion, makes it possible for there to occur changes in a pattern of material extensional organization at different parts of the continuing-existence of that pattern. Change in pattern of organization occurs with, requires, the change of noncoexistent-sequential-difference.

Change in extensional relations between the components of a fourth level pattern constitutes a change in the self-identity of the pattern. Change in self-identity can occur only at different parts of noncoexistent-sequential-difference. It requires the change of noncoexistent-sequential-difference, that of both motion and continuing-existence. This is a stage in the development of change of self-identity, a development from the change of self-identity of ongoing motion from one part of the motion to the following noncoexistent part, to the change of the self-identity of extensional pattern of organization of a moving unit to space, to this stage, the change of self-identity of the extensional pattern of two coexistent units of matter.

Change in self-identity of the extensional pattern of relations between a moving unit and a static unit is the fourth form in which the stages of the development of change in an aspect of self-identity occur. This form is, Change from one individually distinct pattern of extensional relations between a moving unit and a static unit to a following noncoexistent individually distinct pattern—change of relations of materially differentiated extensional pattern. Emergence is a process of change in self-identity, a process of the creation of new self-identity. Emergence is a process of creation.

The change in extensional relations between a moving unit and a nonmoving unit is the change whereby there occurs new pattern of organization of a group of units. The motion that initiates these changes in extensional relations is a case of noncoexistent-sequential-difference, and the newness that occurs with these changes is sequential enhancement. The changes in extensional relations between the units are all cases of noncoexistent-sequential-difference. The new direction, distance, and positional orientation relations, the new pattern of organization, are all cases of sequential enhancement. The development-of-origin of emergence is a process of change in self-identity of a pattern of material organization, which occurs by way of the sequential enhancement of change development. At its origin, emergence is a process of creation by way of sequential enhancement.

Why Is There Emergence?

In short:

◊ There is emergence because existence initiates existence.

- Because existence initiates continuing-existence;
- Because with continuing-existence, the existence and nature of what goes before determines by way of determinate consequent-existence the existence and nature of what follows;
- Because continuing-existence is change, and;
- Because the change of continuing-existence provides a context for the existence of change in relation to that which continues to exist.

◊ There is emergence because coexistent matter and space each initiates its own continuing-existence.

- And because the change of the continuing-existence of matter and the change of the continuing-existence of space together provide the context for the existence of change in relations between matter and space.

◊ There is emergence because motion initiates continuing motion.
- Because continuing motion is change;
- Because the change that is motion initiates change in the extensional relations between matter and space, and;
- Because change in extensional relations of a unit of matter to space constitutes a change in the pattern of relations between the unit and spatial place.

◊ There is emergence because change in extensional relations of a unit of matter to spatial place constitutes change in extensional relations to other units of matter occupying spatial place.
- Because change in extensional relations between units of matter constitutes change in the pattern of organization of those units;
- And because change in the pattern of organization of units of matter constitutes the emergence of a newly existing pattern of the units, and;
- Emergence is the universal determinate process of creative change based on consequent-existence by which newly occurring patterns of material organization come into existence.

If you want to understand why something exists, follow the development.

Chapter 5

The Origin of Cause

The identification of the origin of emergence is but the initial step of providing a description of this process of creation. Of special interest is the relation of the nature and roles of emergence to the nature and roles of cause. The developments-of-origin of both emergence and cause are stages in the development of determinate-reality. Both are determinate processes. Both are stages in the development of consequent-existence. Emergence, however, is developmentally prior to cause. In the development of determinate-reality and consequent-existence, the development-of-origin of emergence is prior to the development-of-origin of cause. The development-of-origin of emergence occurs with the motion of one unit in the presence of another unit, a nonadjacent relation. That of cause occurs with the collision situation, an adjacent relation. The causal relation is an emergent factor that results in a development to another stage of emergence. Determinate-reality, consequent-existence, and emergence all develop with the emergence of cause.

All that is required for the development-of-origin of emergence is that extensional relations change between units of matter, with the consequent occurrence of change in pattern of extensional relations, the emergence of new pattern of material organization. It does not matter what kind of change in extensional relations, just that there is change. At the development-of-origin of cause, it matters what kind of change in extensional relations is occurring just prior. This chapter, then, presents a description of developments leading up to, and including, the development-of-origin of cause, providing an understanding of what cause is, foundationally, and why it exists as a factor of reality.

The Development That Leads to Cause
A Review of the Development of Initiation

Cause is a form of initiation, and at its development-of-origin, it constitutes a major stage in the development of initiation.

Reality is constantly developing. What it is that initiates a development, or what it is that occurs in a situation and in association with an initiator sets the nature of a development, changes as development progresses. Foundationally, development occurs by way of the initiators, space, matter, and motion, while at developed stages it is the associated factors that play the roles that make one stage of development distinct from the previous stage. The transition from initiator characterized forms of development to forms that derive their nature from the associated factors takes place in the early stages of development.

With the form of development that is continuing-existence, mere existence initiates the development. In the simplest case, the foundational case—the continuing-existence of static space—it is the continuing-existence of something that is changeless, and there is here only one aspect, type, or form of development. Matter continuing-existence occurs with the existential context of continuing-existence provided by space, and the nature of the development that occurs with matter continuing-existence is identical to the nature of the development that occurs with spatial continuing-existence, except that it is matter that is continuing to exist. No other factors play a role in the nature of the development.

In the more developed situation of a unit of matter moving in space, there are two more forms of intrinsic development initiated by the motion—the ongoing motion, and the continuing-existence of the motion. Motion continuing-existence, like that of matter, is identical to that of space, and again there are no other factors playing a role. With each of the three cases of continuing-existence, the development is entirely intrinsic to the initiator.

Chapter 5: The Origin of Cause

With space what continues to exist is itself changeless, but with motion what continues to exist is a form of change. When motion occurs, change occurs, development occurs. Motion initiates this change, this development, by continuing to exist, by continuing to be what it is. Since motion initiates this development simply by being what it is, with no other factor playing a role in the nature of the development, the development is again entirely intrinsic to the initiator. So far there are two types of factors that initiate development—existence, for continuing-existence, and motion, for ongoing motion.

Initiation next becomes a little more involved with the occurrence of initiation situation. The first development that occurs by way of initiation situation is change in extensional relations of a moving unit to spatial place. To have an extensional relation, there must be in the situation a factor other than the moving unit, in this case spatial place. In this form of development, then, the nature of the development is in part due to a factor that occurs in association with the initiator. Because extensional relations are relations with something extrinsic to the moving unit, those relations are extrinsic, and, thereby, so also are the developments that go with changes in those relations. With the initiators themselves, the developments are intrinsic to the initiators. With initiation situations—initiators plus associated factor—the developments are extrinsic to the initiators, albeit intrinsic to the situation.

The development-of-origin for initiation situation is the development from the situation of one unit stationary in space, with its unchanging extensional relations with spatial place (a situation wherein the only initiator roles are those of continuing-existence), to the situation of a unit moving through space, with its changing extensional relations (a situation wherein there are three more initiator roles, one of which is a role for the initiator motion in association with factors of a larger context). It is with the development-of-origin of initiation situation that the transition occurs from initiator characterized forms of development to forms that derive their nature from the associated factors. And the first stage of initiation that gets its character from associated factors is the first stage of organizational initiation, which occurs at the third level of extensional pattern.

The General Pattern of the Development

Both emergence and cause originate at the fourth level of extensional pattern, pattern differentiated by units of matter. And both emergence and cause are forms of initiation. The forms and developments of initiation that occur at the third level of extensional pattern provide the foundation and set the nature of the forms and developments of initiation at the fourth level. At the third level, there are three stages of the development of initiation that are of particular significance for the later origin of cause.

What distinguishes the three stages at the third level is the direction of the motion in relation to any particular spatial place. At the fourth level, it is the direction of the motion of one unit in relation to another, stationary, unit. These stages which occur at both levels are, (1) motion directly away from a specific spatial place, or directly away from the stationary unit, (2) motion passing-by a spatial place or stationary unit, and (3) motion directly at the spatial place or unit (Figure 5.1).

The general pattern of development, then, is three stages of the development of initiation at the third level of extensional pattern, then three equivalent stages at the

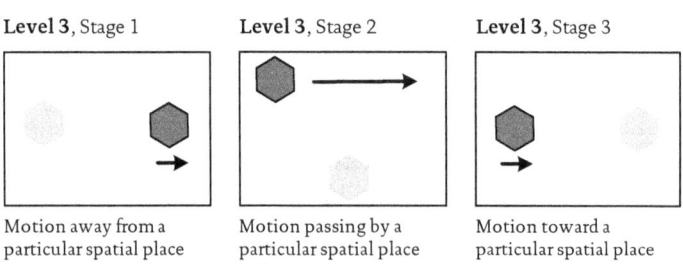

Figure 5.1 *Three stages of initiation at the third level of extensional pattern.*

Chapter 5: The Origin of Cause

fourth level of extensional pattern. It is at the end of the sixth stage of this sequence, at the end of the third stage of the fourth level, that cause originates. It does so at that point because it is there that substantiality first begins to play roles in initiation other than that which differentiates pattern or that which moves.

Factors That Determine This General Pattern of Development

What determines the sequence of development from one level of extensional pattern to the next is, of course, the role of the additional unit. At the third level there is the single unit with its pattern of extensional relations with all spatial places. At the fourth level there are the two units, each with its individual third level pattern of extensional relations with spatial places, and in addition there is also the fourth level pattern of extensional relations between the units themselves (Figure 5.2).

It is the direction of the motion in relation to a spatial place or to another unit that distinguishes the three stages of development of initiation at the two levels. In each case, the particular set of organizational relations between the moving unit and a spatial place or the other unit results in different consequences. All six cases begin as first stage organizational initiation—the initiation of changes in extensional relations between a moving unit and spatial place. In two of the stages, the existential-pathway-development goes on uniformly, without the occurrence

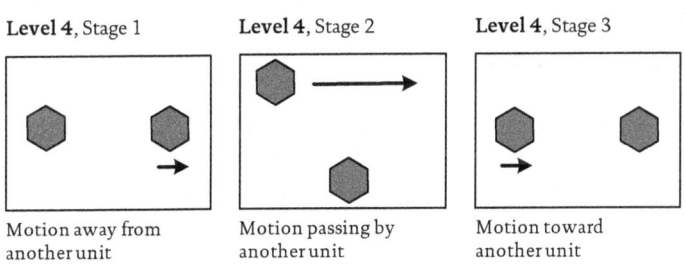

Figure 5.2 *Three stages of initiation at the fourth level of extensional pattern.*

of additional forms of change. In the other four stages, the existential-pathway-development leads into abrupt occurrence of change. It is the factors that distinguish these cases of abrupt change and the consequent differences in the individual existential-pathway-developments of the three stages that set the sequence of development of the three stages at the two levels.

In each case there are the five basic extensional relations that might or might not change. The composite relation, pattern of extensional relations, changes in all three cases since it changes with any motion in relation to any spatial place. The other four extensional relations are the individual relations of distance, direction, positional orientation, and occupation, and which of them change depends on the direction of the motion and the specific spatial place.

Development at Third Level Extensional Pattern

The First Stage of Initiation at the Third Level of Extensional Pattern

The situation of a unit moving directly away from a particular spatial place is the first stage of organizational initiation in that it is the simplest. The only individual extensional relation that changes is the distance of the unit from the spatial place, which continuously increases in a uniform manner without limit. This existential-pathway-development of increasing distance continues without change to its basic nature. The ongoing development of the situation does not lead to change of any type in any factor other than distance, and the change in pattern of extensional relations.

The Second Stage of Initiation at the Third Level of Extensional Pattern

The second stage is motion passing-by a spatial place. Here distance, direction, and positional orientation relations change, which together constitute change in pattern of extensional relations. In this case, however, the existential-pathway-development leads to changes in the nature of the situation, to changes in the manner in which these extensional relations change. The distance relation decreases until reaching a minimum but still existing quantity, then increases without limit. The quantity of change of the direction relation, per specific quantity of motion, increases until the distance relation reaches its minimum, then decreases per specific quantity of motion without further transformation. The rate of change of the orientation relation also increases until the minimum distance between the spatial place and the unit, and then decreases without further transformation. Because it is a composite of the individual extensional relations, the pattern of extensional relations also undergoes a change in its manner of change

Level 3, Stage 2

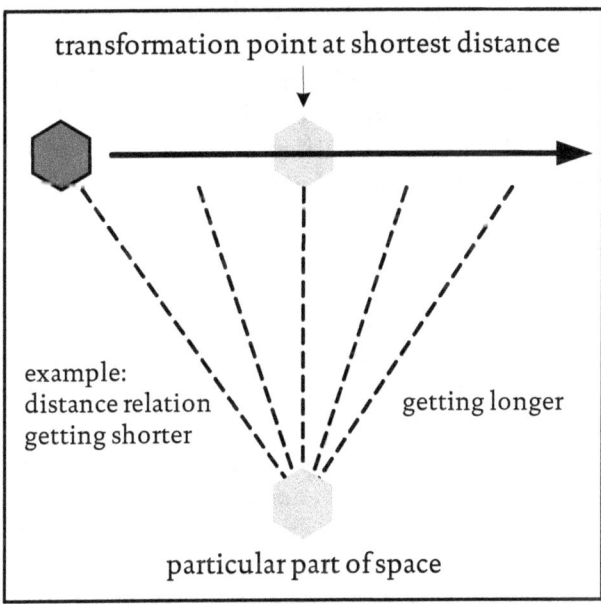

Figure 5.3 *The second stage of organizational initiation at the third level of extensional pattern is passing-by-transformation-point-initiation, wherein there occur abrupt changes in the extensional relations between a moving unit of matter and a particular part of space.*

when the distance relation is at its minimum. With motion past a spatial place, at the point of minimum distance, there are these abrupt changes in the way the extensional relations are changing. The point of minimum distance is a transformation point (Figure 5.3).

Prior to the stage of organizational initiation, all cases of change—three cases of continuing-existence and motion—have been uniform, without variation in the manner of change. In these cases, there is a uniform initiation of the changes, and any part of an ongoing change is isomorphic to the part that occurred immediately prior and to the part that occurs immediately after. The consequences are uniform and isomorphic because they are intrinsic to the initiators.

With the first case of organizational initiation—motion directly away from a spatial place—the change again has a uniform aspect, but this uniformity occurs in association with a form of variation in the manner of change. The distance relation is uniformly increasing. The uniform aspect is a direct consequence of the initiator, motion. It is still a case of continuous uniform initiation of change. The variation in the manner of change, the increasing aspect, is a consequence of organizational initiation, of the initiator occurring in relation to extrinsic factors and the associated organizational relations. A consequence of the specific factors of this case of organizational initiation is that the parts of the ongoing change are no longer isomorphic. There is a development here from uniform change with isomorphic parts of the consequence (continuing-existence and motion) to uniform change with nonisomorphic parts of the consequence (increasing distance relation). In all stages of initiation to here, however, the change occurs in a continuous uniform manner, each in its own way.

With the second stage of organizational initiation—motion passing-by a spatial place—the existential-pathway-development at first progresses in a manner similar to that of the first case, uniform initiation of various ongoing extensional changes each of which is undergoing a uniformly occurring variation in its manner of change. But only to a point, then abrupt change occurs. It is still uniform initiation of change by motion, all the way through. There are various, uniformly occurring, progressive nonisomorphic changes, which are there on both sides of the transformation point. But, because of the abrupt changes due to the role of the transformation point, there is an additional nonisomorphic aspect. Those uniformly occurring, progressive nonisomorphic changes are different on the two sides of the transformation point, different in orientation or different in their aspect of changing quantity. This situation, motion in association with a transformation point, initiates additional aspects of nonisomorphic change, and is a developed stage of organizational initiation.

The development to motion in association with a transformation point is particularly important in the development of reality. It is the development-of-origin of developmental transformation point and existential-pathway-transformational-development. The foundation of existential-pathway-development, with change development, is the continuing-existence of space. It develops then, with the addition of the continuing-existence of matter. Change development existential-pathway-development again develops with the addition of motion, both with ongoing motion itself, and with the continuing-existence of motion. This situation can be stable, as exemplified by the case where the unit moves away from a spatial place without limit and without transformation of the nature of any aspect. However, with motion passing a spatial place, the situation is stable only to a point, whereat transformation of the manner of change of distance, direction, and positional orientation relations occurs, and thereby transformation occurs in the manner of change of pattern of extensional relations. This point in the simultaneous continuing-existences of the components of the situation, this point in the motion of the unit, this point in the development of the situation where the abrupt change occurs, is the developmental transformation point, and situations where these transformation points occur are cases of existential-pathway-transformational-development. A developmental transformation point is a point where change occurs due solely to the ongoing development of the situation without the addition of other factors. The second stage of organizational initiation at the third level of extensional pattern is passing-by-transformation-point-initiation.

Existential-pathway-transformational-development is a developed but still somewhat foundational origin of change. It is a developed stage of organizational initiation—transformation point initiation. Developmental transformation points and existential-pathway-transformational-development are significant for their roles as a foundational source of change, for their roles in the early development of emergence, and as required factors in threshold situations. Threshold situations are initiation situations wherein an initiator, motion, in association with the organization of the components of the situation (organizational initiation), reaches a point (developmental transformation point) in the ongoing existential-pathway-development of the interrelating components such that additional change occurs in the situation (transformation point initiation, existential-pathway-transformational-development). An example of threshold and the role of a transformation point is the point in the existential-pathway-development of a filling reservoir when the water begins to pass over the spillway.

The Third Stage of Initiation at the Third Level of Extensional Pattern

The third stage is motion directly at the spatial place. This case is, at least at first, similar to the case of motion directly away in that the only individual extensional factor to change is the distance relation. However, in this case the distance is decreasing, and this factor can decrease only to nonexistence. Here again a developmental transformation point occurs.

Chapter 5: The Origin of Cause

Level 3, Stage 3

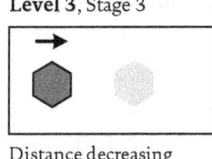

Distance decreasing
Direction unchanging
Positional orientation unchanging

Level 3, Stage 3

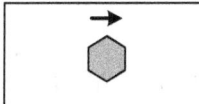

No distance relation
No direction relation
Occupation relation occurs

Level 3, Stage 3

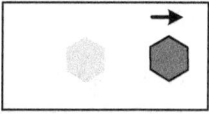

No occupation relation
Distance increasing
Direction unchanging, but reversed
Positional orientation unchanging, but reversed

Figure 5.4 *Passing-through-transformation-point-initiation happens at the third level of extensional pattern.*

As the unit moves along it occupies the spatial place, and at this point in the existential-pathway-development of the situation, the unit acquires an occupation relation with the spatial place, and the distance and direction relations no longer exist. The passing instant of the occupation relation is the developmental transformation point. Past the transformation point the occupation relation no longer exists, and distance and direction relations reoccur. But now the distance relation is increasing. It will do so without limit, and without further developmental change other than that increase. In this aspect the third case is now similar to both the other cases, in that they all result in unlimited increase of distance and no further developmental transformation. The direction relation is also changed. Past the transformation point, the direction is opposite to that which occurred before. The orientation relation has an abrupt transformation. Beyond the transformation point it has the opposite orientation to what it had prior to reaching the spatial place. The opposite side of the unit is now facing the spatial place. And, like the case of passing-by motion, the manner of change of pattern of extensional relations has been transformed (Figure 5.4).

Due to its particular mix of organizational factors, the situation has undergone a one time only existential-pathway-transformational-development. The third stage of organizational initiation at the third level of extensional pattern is passing-through-transformation-point-initiation.

Additional Points of Interest about Transformation Points

The occurrence of specific aspects of change in extensional relations due to the role of a developmental transformation point is a development of initiation. Initiation by way of the motion of a unit of matter in relation to spatial place and in relation to a transformation point, or any developed form of initiation that is the consequence of the role of a transformation point, is transformation point initiation. The form of initiation situation prior to that containing a transformation point was the case of organizational initiation wherein the initiator motion, in relation to the organizational factors of spatial place, resulted in changes in the extensional relations of the moving unit to spatial place. With the role of a transformation point, there are now changes in those changes, changes to the manner in which changes in extensional relations take place. The transformation points are the associated factors that give these stages of initiation their character.

Existential-pathway-transformational-development is an initiation situation, organizational initiation, transformation point initiation, which occurs as the result of the organizational factors of an ongoing change leading inevitably to a point where another aspect of change occurs in the development of the situation. No additional factors joining the situation from outside are required. The transformations are consequences of nothing other than the intrinsic natures of the components of the situation and the factors of their organization.

Besides initiation, there are with these first two cases of existential-pathway-transformational-development, developments of seven other factors of change development. These factors are:

1. Change;
2. Determinate-reality;

3. Consequent-existence;
4. Existential-dependency;
5. Change in aspect of self-identity;
6. Sequential enhancement, and;
7. Pattern of noncoexistent-sequential-difference.

Because transformation points differentiate unique points in ongoing existential-pathway-development, they differentiate unique points in the various forms of noncoexistent continuance that play roles in the existential-pathway-development of change development. Transformation points differentiate unique points in ongoing motion. They differentiate unique points in continuing-existence, in continuance-of-being. They mark unique points in the eternal development of reality.

Existential-pathway-transformational-development is important for the origin of cause because that origin occurs at a transformation point, that is, cause originates at a point where change occurs due solely to the ongoing development of the situation without the addition of other factors.

Development at Fourth Level Extensional Pattern

Because space provides the existential context for all else that exists, the extensional relations between spatial places are the foundation for the extensional relations of higher level extensional pattern. To exist at all, higher level extensional pattern must conform to spatial extensional pattern. The extensional relations of a unit of matter with spatial places, the third level of extensional pattern, are those of the spatial place the unit occupies. At the fourth level of extensional pattern, the extensional relations between two units are the same as the relations between the two spatial places the units occupy.

Changes in extensional relations that occur at the third level due to motion must also conform to the foundational relations between spatial places. At the fourth level, changes in these relations between units of matter are based on the equivalent changes of the third level. Thus the existential-pathway-developments of changing extensional relations of the higher level are based on those of the lower level, and the existential-pathway-transformational-development of the third level sets the pattern of that form of development at the fourth level. The first two of the three stages of the development of initiation at the fourth level are essentially equivalent to those stages at the third level. The occupation of the transformation point of the third stage by a unit of matter radically alters development at that point.

The First Stage of Initiation at the Fourth Level of Extensional Pattern

At the third level when a unit moves directly away from a spatial place, no transformational development occurs (Figure 5.1, Stage 1). The same set of relations hold for the situation in which a unit moves directly away from another unit which is occupying a specific spatial place (Figure 5.2, Stage 1). It makes no difference, in what happens in the existential-pathway-development of the situation, that a unit is moving away from a spatial place that is occupied by another unit. The role of the stationary unit is only the differentiation of fourth level pattern such that the moving unit has increasing distance relation with both the spatial place and another unit. At the fourth level, the one unit could, in an otherwise empty universe, continue moving away from the stationary unit with ever increasing distance, never undergoing any further change than the uniformly changing distance. This is simple fourth level organizational initiation. The existential-pathway-development of this situation does not lead to more forms of change unless other factors enter into the situation. If left to itself, this group of two units would continue dispersing infinitely, the situation remaining fundamental fourth level organizational initiation.

At the fourth level, there are several mixes of factors, which are simply the motion of one unit in the pres-

Chapter 5: The Origin of Cause

ence of a stationary unit, that are organizationally different but which are all cases of the development-of-origin of emergence. The three stages of the development of organizational initiation that are being followed at this fourth level of extensional pattern are each a development-of-origin for emergence. Emergence is an initiation situation. With organizational initiation at the fourth level, new pattern of material organization comes into existence. Whether the one unit is moving directly away from the stationary unit, is passing-by, or is heading directly at it, extensional relation between the units changes, and with any such change there is change in pattern of extensional relations, the occurrence of new pattern. This form of emergence is characterized by the continuous uniform manner in which new pattern of material organization emerges.

Because the existential-pathway-development of the first stage of organizational initiation at the fourth level continues on without further intrinsically originated change, there is no further development of emergence. The simplest case of the origin of emergence goes with the simplest case of fourth level organizational initiation. Like the development of organizational initiation at this level, with the cases and stages of emergence it makes a difference in which direction the unit is moving.

The Second Stage of Initiation at the Fourth Level of Extensional Pattern

At the third level, when a unit moves past a spatial place, the unit passes through a transformation point, and various changes occur in the manner of change of certain extensional relations. Again, the equivalent set of relations and changes occur at the fourth level when a unit moves past another unit occupying a specific spatial place. That is motion passing-by—any direction of motion of the unit that takes it on a path toward the static unit, but misses it, and goes on by. Here also, it makes no difference in what happens with the existential-pathway-development of the situation that the place being passed by is occupied.

Origins of Self-Organization, Emergence and Cause

Level 4, Stage 2

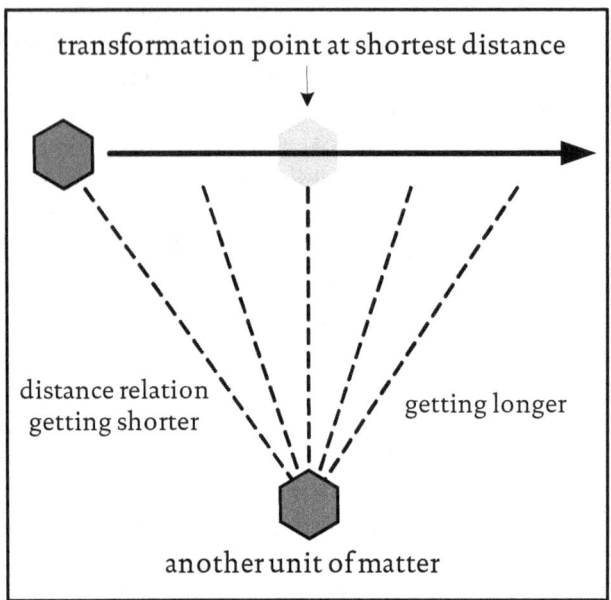

Figure 5.5 *Passing-by-transformation-point-initiation also occurs as the second stage of organizational initiation at the fourth level of extensional pattern, where there are abrupt changes in the extensional relations between a moving unit of matter and a stationary unit.*

The stationary unit just differentiates fourth level pattern such that the moving unit has changing extensional relations with both the spatial place and another unit. Like the spatial place it occupies, the stationary unit plays no role in the abrupt changes in those relations initiated by the transformation point and the motion, other than differentiating the fourth level pattern the extensional relations of which get abruptly changed along with those of the third level. Because the stationary unit does not play a role in changing the nature of the ongoing development of the situation, the fourth level follows the pattern of change set by the third level. It is a one time only transformation and the one unit continues to move away from the other without limit, as long as no extrinsic factors play a role. This

Chapter 5: The Origin of Cause

is fourth level, passing-by transformation point initiation, the second stage of organizational initiation at this level (Figure 5.5).

The existential-pathway-development of motion passing-by leads to abrupt change in the manner of change of the distance, direction, and positional orientation relations between the two units. There occurs a change in the manner in which organizational initiation takes place. There is another factor associated with the initiator motion, a transformation point. This factor plays a one time only role.

Before the transformation point, it is second stage fundamental organizational initiation, and it is so at both the third and the fourth levels. There is a continuous uniform initiation of change in the three extensional relations. After the transformation point it is again second stage fundamental organizational initiation, with continuous uniform change in the same three relations, albeit with certain alterations in the manner in which they change. At the transformation point, however, it is organizational initiation in the form of transformation point initiation, a developed form. With the existential-pathway-development of the situation, it is fundamental organizational initiation before the point of minimum distance, it is transformation point initiation at that point, and then the fundamental form reoccurs, although altered, after the unit passes beyond the point of minimum distance. A form of a factor, organizational initiation, occurs at an earlier part of an existential-pathway-development, develops to another form, then reoccurs, at a later part of the situation development, as the original form, somewhat altered.

Since at third level pattern any motion is motion directly away from, passing-by, or directly at an infinity of spatial places, and since at fourth level pattern there are many, probably an infinity, of units of matter, such that any motion is motion directly away from, passing-by, or directly at some unit or other, the development-of-origin for any of these six stages of organizational initiation is the development that adds motion, any motion, at either level. All three cases, at each level, have the same development-of-origin.

With the organizational initiation that occurs with motion directly away from a spatial place or from another unit, there is change in only the one extensional relation, distance. With motion passing-by there is change in three. So, even though the development-of-origin for these two cases of organizational initiation is the same either at the third level or at the fourth level, the one case is more complex than the other.

As emergence develops, a great many different types of initiation situation serve as the process by which emergent creation occurs. The differences between the various cases of organizational initiation at the fourth level are the differences there between the cases of emergence. Fourth level, fundamental organizational initiation is the process of emergence at its development-of-origin. Each of the three stages at this level begins with some form of fundamental organizational initiation, with the fundamental form of emergence. Then, the development of organizational initiation that occurs in the form of passing-by transformation point initiation is also a development of emergence, a form of emergence characterized by the role of a transformation point, and thus by abrupt change in the manner in which new pattern of material organization emerges. Because the existential-pathway-development of the second stage of organizational initiation at the fourth level continues on without further intrinsically originated change, there is again in this case no further development of emergence.

The Third Stage of Initiation at the Fourth Level of Extensional Pattern—And the Emergence of Contact

At the third level, when a unit moves directly at a spatial place, the distance relation decreases until it no longer exists, the spatial place acquires a role as a transformation point as the unit passes through, and various changes oc-

Chapter 5: The Origin of Cause

cur in the extensional relations between the unit and the spatial place. At fourth level extensional pattern, the specific place and the transformation point again coincide, but here they are occupied by the stationary unit. The moving unit cannot simply pass through as with an unoccupied place. At the third level, the existential-pathway-development of the organizational initiation of motion directly at the spatial place results in the unit passing through the place, with the consequent occurrence of passing-through-transformation-point-initiation. At the fourth level, because there is a unit occupying the transformation point, and because the existential-pathway-development of the organizational initiation of motion directly at the stationary unit results in collision, this is collision-transformation-point-initiation. The stationary unit plays a role here in transformation point initiation that gives this case its character, and that makes this case a new stage in the development of transformation point initiation.

Collision, then, is an initiation situation, organizational initiation, a developed case of organizational initiation, a case of transformation point initiation, and again a developed case—collision-transformation-point-initiation. Initiation situation is a situation wherein an initiator occurs in association with another factor such that there are consequences extrinsic to the initiator. Organizational initiation is a type of initiation situation wherein,

Level 4, Stage 3

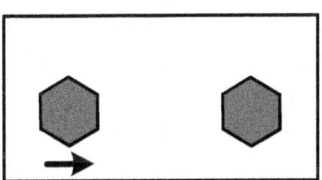

Motion toward a stationary unit of matter

Level 4, Stage 3

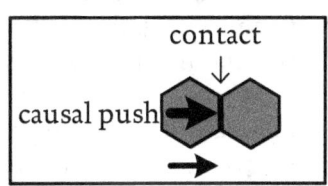

Collision-transformation-point-initiation

Figure 5.6 *Collision-transformation-point-initiation. Abrupt changes in extensional relations between the moving unit and the stationary unit, and the emergence of contact and cause.*

foundationally, the motion of matter occurs in association with spatial place such that there are consequent changes in the extrinsic extensional relations between the moving unit and spatial places. Transformation point initiation is a type of organizational initiation wherein, foundationally, the motion of a unit occurs in relation to a particular spatial place such that the role a transformation point plays results in abrupt changes in the extensional relations between the unit and the particular spatial place. Collision-transformation-point-initiation is a type of transformation point initiation wherein the motion of a unit occurs in relation to a particular spatial place that is occupied by a stationary unit such that the combined transformation point roles of that place and the stationary unit result in abrupt changes in the extensional relations between the moving unit and both the spatial place and the stationary unit, and result also in the emergence of cause (Figure 5.6).

The emergence of cause is a two part process in that there are two stages in a collision situation which are both required for the emergence of cause—the emergence of contact and the emergence of push. The role of the stationary unit in collision-transformation-point-initiation is a role of its substantiality, a role of its primal existential nature, and the two stages of a collision are cases of primal factor initiation. Foundationally, collision-transformation-point-initiation is a case of primal-factor-transformation-point-initiation.

Contact comes first. It is developmentally prior to push, with push existentially-dependent on contact. When the distance between the units becomes nonexistent, there is no immediate change in the direction or positional orientation relations. The nonadjacent relation develops into an adjacent relation. However, matter is not immaterial. Something is there, matter, substantiality—two units of substantiality in adjacent relation. In this situation a new relation emerges—contact. This new relation is existentially-dependent on the adjacent relation. Because the units are there together without any distance between them, without any spatial place between them, they are in contact with one another—because they are substantial.

This is a new role for substantiality. The two previous roles of substantiality, material differentiation of pattern and that which moves, are significant in large part for where they occur, that is, as factors of the existence and change of extensional relations. With the material differentiation of pattern substantiality plays its role because something is-there that has an aspect of its mode-of-being that is more than the immaterial three-dimensional extension of spatial place. Something is-there occupying spatial place, and thereby differentiating a part of it from other parts. By differentiating one part of space from another, substantiality plays a role in differentiating the extensional relations between spatial places and between anything that occupies those places. Which spatial place and which extensional relations are differentiated depends on where the unit of substantiality is.

With the role of substantiality as that which moves, it still plays its role of materially differentiating spatial place and extensional relations. However, the motion of matter changes the extensional relations between the moving unit and spatial place, and whatever occupies space. Again, what is differentiated depends on where the substantiality is occurring, on where it is moving through, as do the specific changes that are occurring.

The new role of substantiality in the emergence of the contact relation is based in large part on what substantiality is. To exist is to have organization, so all the appropriate organizational factors are there playing their roles. If two units come into adjacent relation to one another, that is a relation based on their spatial locations—an organizational relation. If two units come into contact relation with one another, that is a substantiality based relation, a relation in which there is a role that is more than occupation or differentiation of place, or change of occupied place—a supra-organizational relation. Just as extensional relations are based on the nature of immaterial spatial place, supra-organizational relations such as contact are based on the nature of matter.

When distance ceases to play a role, substance begins to play its new role. All the other factors of the previous stage are there and still playing their roles, but now it is this new role of substantiality that gives the situation its character. Substantiality is a primal factor, and in this collision situation, it begins to play roles other than those of differentiating material pattern and being that which moves. Roles are played in the contact relation by the substantiality of both units, but it is the substantiality of the moving unit, playing its role as a factor of an initiator, that has an active role in the origin of a new stage of this development. The substantiality of the moving unit has three roles in the origin of the contact relation, (1) differentiation of material pattern, an organizational role, (2) that which moves, an initiator role, and (3) as the basis of contact, a primal factor role. The two roles of the substantiality of the stationary unit are passive, as a factor of organization and as a primal factor.

This is a case of transformation point initiation in that initiation and a consequence occur at a specific point as a result of the ongoing existential-pathway-development of the existing situation, without the addition of any new factor. At third level extensional pattern, the initiation that occurs at a transformation point is based on extensional relations and the changes that occur there are changes in extensional relations. At fourth level extensional pattern, the initiation that occurs at the transformation point where there is collision is based not only on extensional relations, with changes in extensional relations, but also on substantiality, with change that is existentially based on the intrinsic nature of substantiality.

With the prior nonadjacent situation all the relations were extensional and not direct. The contact relation has an aspect that does not occur with extensional relations, an aspect that involves the direct relation of each

Chapter 5: The Origin of Cause

unit with the other. Contact is interrelational. The developments of initiation and emergence that occur here are due to this interrelational aspect.

Contact emerges in an initiation situation. This situation involves factors of organizational initiation, factors of transformation point initiation, and requires in addition the new supra-organizational role of substantiality as a primal factor, which makes this a stage of primal factor initiation. This role of substantiality occurs when two units of matter are in adjacent relation, making this a developed form, adjacent-substantiality-primal-factor-initiation. Primal factor initiation develops once more in the collision situation

Since initiation is the origin of the change that occurs in the process of emergence, emergence develops with the development of initiation. Here it develops from stages characterized by changes in extensional relations and the emergence of new pattern of material organization, from stages characterized by foundational organizational initiation and passing-by transformation point initiation, where the roles of matter are the nonadjacent and noninterrelational differentiation of pattern, and change of pattern. It develops to a stage that involves changes in extensional relations, the emergence of new pattern of material organization, and organizational initiation, but is now characterized by collision-transformation-point-initiation and adjacent-substantiality-primal-factor-initiation, where the roles of matter are the adjacent and interrelational differentiation of pattern, and as a primal factor such that there occurs the emergence of a supra-organizational relation, contact.

The development from the organizationally characterized origin stages of emergence to the primal factor stages occurs with the new role of substantiality in the contact relation. This is a development where direct relation, interrelation, first plays a role in the process of emergence. It marks a particularly significant stage in the development of emergence because all causally based stages of emergence are existentially-dependent on contact and interrelation. Development of emergence, beginning with contact, involves not just motion and extensional relations, but factors of the intrinsic nature of the units of matter. The consequences of developed stages of emergence result from the manner of togetherness of the components—their manner of reorganization, separating, or combining—plus the roles of the intrinsic nature of the units. With contact substantiality determines the nature of the emergent consequences, but at later stages existential quantity, shape, density, organization of two or more units combined into a single unit, and numerous other factors determine the nature of the emergent consequences.

The emergence of the contact relation is determinate because it is a determinate process, change of extensional relations between units of matter due to motion, that brings it about. The contact relation is characterized by the role of substantiality, which is an intrinsic factor of matter. The emergence of this relation is determinate because an initiator, motion, operates in the context of other factors, two units of matter, their adjacent relation, and the intrinsic nature of substantiality. The existence and nature of the consequence, contact, is determined by the role of the initiator in conjunction with the intrinsic qualities of the other factors, in particular substantiality.

The Development-of-Origin of Cause
The Third Stage of Initiation at the Fourth Level of Extensional Pattern—And the Emergence of Push

Because motion is the initiator, the development of the situation goes with the flow of the motion. The continuing-existences of all the components of the situation are unidirectional, and the motion, having a one on one developmental relation with the three cases of continuing-existence, is itself existentially unidirectional. The development of the collision situation, then, occurs unidirectionally from the moving unit to the stationary unit.

Chapter 5: The Origin of Cause

The stationary unit is in the way of the moving unit, and at the moment of contact it constitutes an obstruction to the motion. The intrinsic nature of motion, however, is to continue—consequent-existence by way of continuing-existence of the intrinsic nature of motion. When the moving unit comes up against the stationary unit, the moving unit will naturally keep on going, and by way of the contact relation, the substantiality of the moving unit will press against the substantiality of the stationary unit. The emergence of this pressing, this push, is the development-of-origin of cause.

With current data and understanding, this is the simplest case. Because why matter exists is not known, and because the nature of the primal form of matter is also not known, we can expect some fascinating surprises that may well alter what is understood to be the simplest or primal case. Perhaps there is a simpler case of the development-of-origin of cause, or this case may be the simplest but yet be a developed stage, with the actual development-of-origin a more complex situation. Because matter moves, and because all identified matter appears to be unitary in one way or another, from planets to quarks, the situation of a moving unit colliding with a stationary unit is the simplest form of the origin of cause that can be currently reached by way of structural logic.

What actually happens when contact is made between two units and push is applied by the one to the other is complex, with various aspects of the nature of substantiality, such as elasticity, playing roles. It is not the intention here to describe all of that complexity, but only to point out the origin of cause, to identify the factors that play roles in that origin, and to identify the consequence of those factors, the intrinsic nature of the foundational form of cause.

A great deal happens at the transformation point that occurs as a consequence of the motion of a unit of matter directly at a stationary unit. It all begins in the form of organizational initiation as the one unit moves toward the other. However, as soon as the distance ceases to exist there is adjacent relation. And as soon as there is adjacent relation there is contact. As soon as there is contact there is push. As soon as there is push there is transference of motion. As soon as there is transference of motion there are two units in motion relative to space, and thus in mutual motion relative to each other, a new, developed, stage of organizational initiation. This all happens almost instantaneously, but in that short moment many factors develop through several stages.

Contact sets the stage for the origin of push. There can be no push if there is no contact. Since push operates by way of contact, cause is directly existentially-dependent on contact. Push, the second supra-organizational factor, is a consequent emergent of a stage of emergence that is characterized by contact, the first of these supra-organizational factors.

An event of initiation is an event of development. Stages of initiation are the developments-of-origin for the stages of other factors. Contact is a stage in the development of the roles of matter. Push, also, is a stage in that sequence of development. Each is the consequent of a distinct stage of initiation. Contact is the consequence of adjacent-substantiality-primal-factor-initiation. And push is the consequence of collision-primal-factor-initiation.

Adjacent-substantiality-primal-factor-initiation is a developed stage of primal factor initiation. It is the situation, the beginning stage of a collision situation (initiation situation), where a moving unit has come into adjacent relation (organizational initiation) with a stationary unit (transformation point initiation), wherein a new role for substantiality has the consequence that interrelation and contact emerge.

Collision-primal-factor-initiation is the next developed stage of primal factor initiation. It is the situation, a subsequent stage of a collision situation (initiation situation), where a moving unit (organizational initiation) is in contact (primal factor initiation) with a stationary unit (transformation point initiation), wherein a new role for substantiality in motion has the consequence that interaction and push emerge.

Once contact is emergent in a situation where motion is initiating collision, it then plays a role in the next stage of initiation in this existential-pathway-development. Since initiation is the source of change in the process of emergence, there are here in a collision situation two stages of emergence, one with contact as an emergent factor and another, based on contact, from which push is emergent.

Again the role of substantiality develops. Just as it is substantiality that exists in a manner different from immaterial space, that occupies spatial place, differentiates material pattern, moves, and contacts, it is substantiality (in motion) that pushes.

The emergence of push, like the emergence of contact, is the emergence of a factor that has an aspect of its nature that is something more than merely organizational. Cause, like contact, is a supra-organizational emergent based on the intrinsic substantiality of matter. However, to exist is to have organization. The causal relation is organized, with roles of varying significance for all the factors of organization that have been mentioned to this point. For example, the factor push is organized in space, in material pattern, in time, in motion, in initiation, in contact, and in its action. Indeed, for cause to emerge, there must be an organizational aspect of isomorphism between two organizational factors—the direction relation between the units, and the direction of the motion. The direction relation between the two units and the isomorphic direction of the motion also set the initial direction of the push. In this manner, cause is a pattern of organization of factors. When substantiality occurs in the context of these organizational factors, the supra-organizational factor push is emergent.

How much motion there is plays a role in how much push there is. The push is a consequence of the moving substantiality, and through that relation the quantity of the push is a consequence of the quantity of the total motion of the pushing unit. There are four roles of quantity that occur with motion. The distance traveled, and the time it took to go that distance, are two of the four quantities of motion. They are the quantities of the history of the motion. The amount or existential quantity of matter that is moving as a unit, and the speed at which it is traveling, are the other two quantities of motion. They are the quantities of the current existence of the motion. The greater the quantity of moving matter, the greater is the quantity of motion. The greater the speed of the moving matter, again, the greater is the quantity of the motion. The quantity of moving matter and the speed at which it is moving, the two together, are equivalent to the total quantity of motion that is occurring in a particular case. If there is more or less material existential quantity in motion, there will be more or less push of the one unit on the other. Likewise, if there is more or less speed, there will again be more or less push.

All the prior stages of determinate-reality occur here. There are the three foundational cases of continuing-existence. There is the continuance of the intrinsic nature of motion, the continuance of motion itself, in relation to space. There are the changes of extensional relations of a moving unit to spatial place, a stage of organizational initiation at the third level of extensional pattern. Then there are the changes of the extensional relations of a moving unit to a stationary unit, which is another stage of organizational initiation, this time at the fourth level of extensional pattern. And, there is the emergence of the contact relation.

The push relation between the moving unit and the stationary unit, the pressing of the substantiality of the one unit against the substantiality of the other unit, is determinate because all these prior stages play roles in this situation. The immediately prior stage of this existential-pathway-development was the consequence of motion bringing together into adjacent relation two cases of substantiality. But the pathway development does not stop there. The motion initiator continues to play its role. The one unit continues to move, initiating, through the contact relation, the pressure on the other unit. The motion is determinate, the consequent pressure is determinate—Determinate consequent-existence of the emergent factor, cause. With the occurrence together of all the factors of this situation, the determinate emergence of pressure, of push, of cause, is inevitable.

Self-organization originates with the origin of time.

Chapter 6

The Origin of Self-Organization

Self-organization, like cause, is another factor of special interest. Because in its more developed forms it is a type of emergence, it is not surprising that an understanding of the origin of emergence provides an understanding of the origins of self-organization. It originates, however, prior to emergence, and does not become a type of emergence until the origin of that factor. Examples of naturally occurring self-organization that are probably of most interest to most people are those of chemistry, geology, biological evolution, the growth of an organism, ecology, and the origins of social organization. These range from the mildly well developed cases of chemistry to the advanced biological cases. Self-organization, however, originates in exceptionally simple form, and develops through many stages, adding numerous factors to its nature, before it approaches the simpler chemical cases. The earliest stages of its development are presented here.

The intent is to explain why self-organization exists, to show why some factors have self-organizational roles, and to show that they have these roles because of what they are. It begins with the character of the initiators, and develops into the character of initiation situations.

Foundationally, there is self-organization (a) because there is consequent-existence, (b) because the existence and nature of what goes before determines the existence and nature of what follows, (c) because factors intrinsic to what goes before can, and usually do, play roles that determine organizational factors of what follows, (d) because there is emergence of new factors, (e) and because emergent factors can play organizing roles in the further existential-pathway-development of the situations from which they have emerged.

While most cases of self-organization about which people have expressed interest are also cases of emergence, both self-organization and emergence can have aspects or cases which are distinct from the other. Because self-organization originates prior to emergence, not all cases of self-organization are cases of emergence, and because some cases of emergence have supra-organizational consequences and because some cases of emergence are processes of disorganization, not all cases of emergence are cases of self-organization.

The Basics

Existence initiates existence. Existence determines existence. What it is that goes before determines what it is that follows. When factors intrinsic to a prior stage of development determine organizational factors intrinsic to the following stage, that is self-organization.

Self-organization progresses by way of existential-pathway-development. Because the intrinsic factors of any prior stage always determine to some extent the factors of the following stage, there is an aspect of self-organization in every situation involving the initiation of intrinsic organization. Even when a factor extrinsic to a situation is the source of organizational change within that situation, it plays its role there interrelationally with the factors intrinsic to the situation, which gives the development an aspect of self-organization. The analysis of merging situations for the roles of self-organization requires careful tracking of the existential-pathway-developments of the two situations, of the manner in which they interrelate after contact and prior to separation, or of the manner in which they combine into a single situation. Like most other factors, self-organization develops, and different stages require different descriptions, depending on what factors give any particular stage or form its distinct character. There is always, though, an aspect of intrinsic factors determining organizational aspects of the consequent. With this process, the consequence has some form of orderly relation between parts. It is self-organization, not disorganization.

The origins of self-organization are primal. Its development follows, for the most part, the general stages of the development of reality described in previous chapters. The foundational occurrence is with the initiation of spatial continuing-existence. Existence initiates the organization of continuing-existence (simply by doing so, by continuing to exist, continuously, sequentially, with consequent organizational before and after relations between parts). The situation is space, the existence of space, which initiates the intrinsic consequence of its own continuing-existence. Spatial continuing-existence has organization, that of noncoexistent-sequential-difference. Organization is initiated, created. An intrinsic factor, the existence of space, initiates spatial continuing-existence, an intrinsic consequence, and thereby creates organization—an intrinsic developmentally prior factor creating an organized intrinsic consequent factor. That is self-organization.

Self-organization develops, by way of nonpathway factor development, through two more stages of the initiation of the organization of continuing-existence, those of matter and motion. The pattern of the initiation of noncoexistent-sequential-difference is simultaneous and isomorphic in all three stages of the development of the initiation of continuing-existence. The pattern of the initiation of the organization of continuing-existence is the same in the three stages. Thus, these three cases of self-organization are of the same form, differing only in what it is that is continuing to exist (Table 6.1).

The initiation of the organization of the noncoexistent-sequential-difference of motion itself is the next development of self-organization. Because the foundational origin of motion is not known, it is equally unknown whether this stage of self-organization is the result of existential-pathway-development or is a case of nonpathway factor development. In the context of the existential-dependency relation of motion with matter, and the incapac-

Origins of Self-Organization, Emergence and Cause

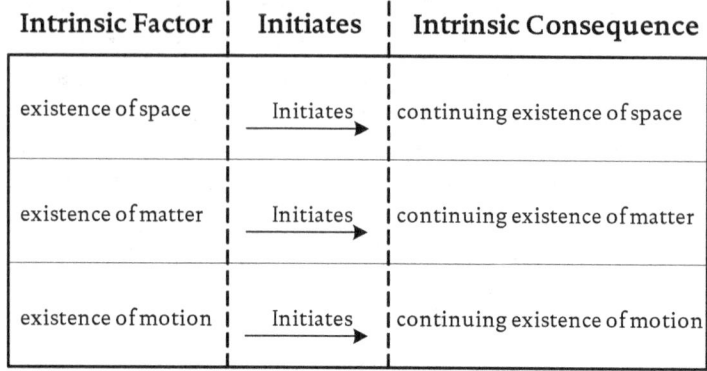

Table 6.1 *The development-of-origin of self-organization occurs with the initiation of continuing-existence, with the occurrence of noncoexistent-sequential-difference, unidirectionality, and sequential before and after relations.*

ity of space to play a role in the origin of motion, the origin of this stage is probably some form of pathway development consequent on the foundational nature of matter.

The initiation of the organization of motion is itself, like all stages and cases of self-organization, an occurrence of existential-pathway-development. Motion, by continuing to be itself, initiates the intrinsic organization of continuing motion. An intrinsic factor, the existence of motion, initiates ongoing motion, an intrinsic consequence, and thereby creates organization—an intrinsic developmentally prior factor creating an organized intrinsic consequent factor. Again, this is self-organization.

Undisturbed motion has the same organization of noncoexistent-sequential-difference as the continuing-existence of motion, which is simultaneous and isomorphic with the continuing-existence of space and that of matter. The organization of motion is simultaneous and isomorphic with the organization of the prior forms of noncoexistent-sequential-difference. Likewise, the self-organization of motion is a different stage and form, but is still simultaneous and isomorphic with the self-organization of the three prior stages.

Chapter 6: The Origin of Self-Organization

The next development is motion in the context of space, motion in relation to spatial place. There are changes in the extensional relations between a moving unit of matter and spatial places. Here self-organization has a different character in which the consequences of the initiation of change are extrinsic to the initiator. All stages to this point are still playing their roles, but now the initiator motion plays its role in the context of the extrinsic factors of spatial place—location, distance, and direction. The intrinsic self-organization of motion itself is occurring in the context of other factors that play roles in the organization of the whole situation. It is an initiation situation, organizational initiation (Figure 4.1).

The situation is a moving unit of matter, space, and the extensional relations between them. A unit has extensional relations with all parts of space, and the motion initiates organizational changes in the distance, direction, and positional orientation relations of the unit with all those parts. Because space is infinite, without boundaries or external factors of any sort, this is a single situation. All factors of this situation are intrinsic to the situation. The motion, an intrinsic factor, initiates new intrinsic organizational relations—self-organization (an infinite case).

Self-organization is a form of change. As such it is always associated with initiation. At its foundational stages it occurs with the primary initiators, the existence of space, matter, and motion, and their initiation of continuing-existence, and with motion, and the initiation of ongoing motion. In each of these stages, the initiator is the situation. All factors of the situation are intrinsic to the initiator, including the consequence and its newly occurring organization. This is not so for the self-organization that goes with initiation situations, beginning with the first stage of organizational initiation (the initiation of changes in extensional relations between a moving unit and spatial places). Here the primary initiator occurs in relation with extrinsic factors that play roles in the process of self-organization, with consequences extrinsic to the initiator. This is a profound transition in the development of self-organization, which sets the core nature of all further developed forms.

The initiation of continuing-existence and the initiation of ongoing motion are determinate, making the associated forms of self-organization determinate. The initiation of changes in extensional relations between a moving unit and spatial places is also determinate, as is the associated self-organization. With continuing-existence and motion, the initiation is determinate because they are cases of the continuance of self-identity. In each case that which exists prior determines that which follows, simply by way of continuing to be itself. With the initiation of changes in extensional relations, the determinate aspect is more involved, because of the roles of the factors extrinsic to the initiator. What those extrinsic factors are, where they are, and when they are determine the what, where, and when of the consequence—the what, where, and when of the self-organized pattern.

At any particular part of the continuing-existence of a unit of matter, it exists at a specific spatial location. The unit shares with the spatial place it occupies the extensional relations that place has with all other spatial places. Motion is sequential, and space has sequential organization of place. When the unit moves, it acquires in a sequential manner the extensional relations of the sequence of spatial place it passes through. With each newly acquired set of extensional relations, the moving unit acquires a different pattern of extensional relations with spatial place. Each of these newly occurring patterns of relations is a unique, newly occurring, pattern of organization. The motion initiates new pattern of organization—the situation is self-organizing. This self-organization is determinate. It is so because the motion is determinate, and because the extrinsic factors of location, distance, and direction are existentially what they are (Figure 4.1).

Motion constantly initiates new part of ongoing motion by way of the determinate consequent-existence

Chapter 6: The Origin of Self-Organization

of the continuing-existence of what it is. This initiation of new part is sequential, each new following part coming immediately after the prior part. Prior parts and their consequent following parts are sequentially adjacent. The prior part develops into the sequentially adjacent following part.

At any particular part of the continuing-existence of motion, that motion exists at a specific spatial location. Because motion is matter passing through space, each following, individually distinct, sequentially adjacent part of motion exists at an individually distinct, sequentially adjacent part of space. Because the sequentiality of motion is a form of determinate consequent-existence, the association of the unit and its motion with a succession of different spatial places is a determinate consequence of the occurrence of the motion in relation to the extrinsic factor spatial place. The determinate nature of the initiator, its intrinsic determinate quality, has the consequence that the changes it initiates with extrinsic factors are also determinate.

When a moving unit is at any particular spatial place, it has that location's extensional relations with the rest of space. When the unit moves on to the adjacent place, it shares the extensional relations of that location. With the sequence of new part of motion goes a sequence of newly occupied spatial places. With the sequence of occupied places goes a sequence of newly acquired extensional relations. The parts of the motion are adjacent. The occupied places are adjacent. And the sequential acquisition of extensional relations is of a sequence of adjacent relations—the occupied place adjacent to the prior occupied place, the distance to any specific location adjacent to the prior distance to that location, and the direction adjacent to the prior direction. Existential-pathway-development goes sequentially from part to part, from part to adjacent part, from stage to sequentially adjacent stage. This is a core feature of self-organizing situations.

The nature of self-organization develops when transformation points play roles. Transformation points (and thresholds, their developed forms) are good examples of the significance of development to a sequentially adjacent part, step, or stage. At this stage, self-organization is of the same form as that when motion occurs in relation to all spatial place, but this stage involves, in each case, an individual location that plays a unique role in organizational aspects of the consequent situation. Due to the roles of transformation points, which are factors intrinsic to the situations, there are organizational changes in rates of change, and in directional and positional orientations—self-organization.

When a moving unit arrives at a transformation point, the foundational form of organizational initiation develops to another form, transformation point initiation. One of the first extensional relations to change abruptly at a transformation point is adjacent relation—it comes into existence (Figure 6.1). This is adjacent relation at third level pattern, that between a spatial place, the transformation point, and a unit of matter. With the adjacent relation that

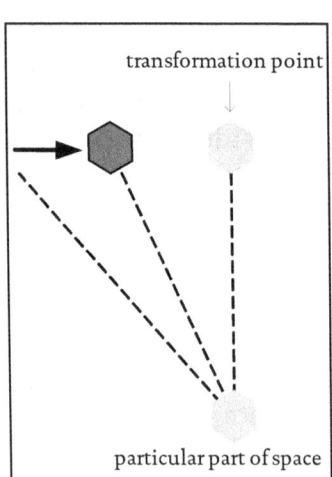 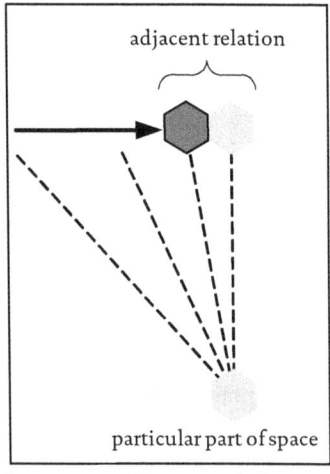

Figure 6.1 *With motion passing by a particular spatial place the first emergent stage of transformation point initiation is adjacent relation.*

Chapter 6: The Origin of Self-Organization

occurs with passing-through-transformation-point-initiation, there are two new organizational relations the later development of which, in the collision situation, will help track the development of self-organization through that more developed situation.

A new adjacent relation occurs in the context of the already existing first level adjacent relation between the spatial place that is the transformation point and the spatial place adjacent to it. The first level relation has two organizational relations that are of interest here. There is the directional relation between the components, between the immaterial spatial places. This is the directional alignment of the adjacent relation in space. And there is the positional orientation relation between the components—what part of one place is adjacent to what part of the other place.

A unit's third level pattern extensional relations with spatial places are the same as the first level pattern extensional relations of the place it occupies. The third level adjacent relation occurs in the place that is the first level adjacent relation, and has the same organizational relations. There is the third level directional alignment between the transformation point place and the adjacent unit just as there is the first level directional alignment between the transformation point place and the adjacent spatial place (occupied by the unit). And there is the positional orientation relation between the transformation point place and the unit, with a specific part of that place adjacent to a specific part of the unit.

Because the third level adjacent relation is newly occurring, with the arrival of the moving unit, the qualities based on that relation are also newly occurring. The directional alignment and the positional orientation of the third level adjacent relation are newly occurring, as a consequence of the intrinsic existential-pathway-development of the situation—as a consequence of self-organization. These two organizational relations develop further through the stages of contact and push. Because they are organizational relations, these developments are cases of the development of self-organization.

Because this newly occurring extensional relation is between a unit of matter and a spatial place, it is a consequence of the first form of organizational initiation, the form at third level pattern of organization. Because the transformation point constitutes one half of the relation, the occurrence of this adjacent relation is the first part of the role of the transformation point, the first part of third level   passing-through-transformation-point-initiation. The self-organization that occurs here does so by way of (a) the simple organizational initiation of motion changing the extensional relations between a unit and a particular spatial place, with decreasing distance relation until there is no longer any distance between the unit and the particular place when the unit arrives at the adjacent place, where (b) transformation point initiation begins to play a role in the self-organization of the situation when the particular spatial place abruptly becomes the place adjacent to the unit.

The next development of self-organization goes with the motion of a unit in relation to a stationary unit. The same set of relations occurs, as with motion in relation to spatial place, except they are now in relation to the stationary unit. Fourth level pattern and the events that occur there are existentially-dependent on third level pattern and what happens there. It is self-organization because the motion of the one unit initiates organizational changes in the extensional relations between the two units, the existential-pathway-developments of the two units in relation constituting a situation. New pattern of extensional relations occurs, and that is emergence. Here the development of self-organization and emergence combine. Here also is the development-of-origin of another core factor of more advanced forms of self-organization. Motion changes the organization of matter. This factor will develop in the collision situation, where, in association with other factors,

Chapter 6: The Origin of Self-Organization

1. Adjacent relation

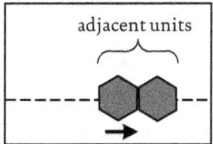

Substantiality based directional alignment and orientation relations

2. Contact relation

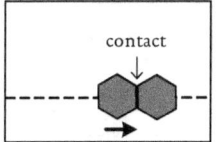

Directional alignment and orientation of the contact relation

3. Blocking

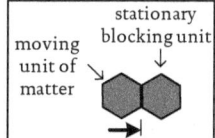

Intrinsic alignment of the blocking event

4. Changed motion of the moving unit

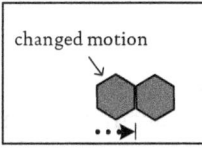

Organizational aspects of direction and parts of motion

5. Push

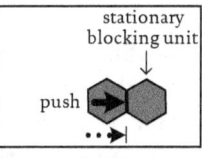

Directional alignment and unidirectionality of push

6. New motion of the blocking unit

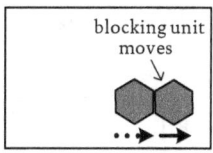

Organizational aspects of direction and parts of new motion

Figure 6.2 *Collision situation has six stages of development that have emergent organizational aspects due to self-organization.*

it acquires a role that is even more critical to the nature of advanced self-organization.

The developments of self-organization at the fourth level of pattern follow those of the third level—except of course for the consequences of motion directly at the occupied transformation point.

Self-Organization through the Collision Situation

The collision situation has six stages of development that have aspects of self-organization—the origins of (1) adjacent relation, (2) contact relation, (3) blocking, (4) the changed motion of the moving unit, (5) push, (6) new motion of the blocking unit. Each of these stages has emergent factors, emergent organizational factors, consequences of self-organization, and three have emergent supra-organizational factors (contact, push, new motion), consequences of emergence only (Figure 6.2). All

Origins of Self-Organization, Emergence and Cause

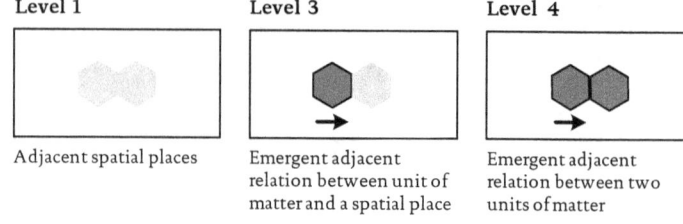

Figure 6.3 *Levels of adjacent relations.*

these emergent factors play roles in one way or another in the succeeding stages of self-organization that occur in the collision situation.

Adjacent Relation

In the collision situation, the first emergent is adjacent relation. There are here two new cases of adjacent relation, one between the moving unit and the place occupied by the stationary unit, the transformation point, and one between the two units. The one between the moving unit and the transformation point is a third level relation, while that between the units is a relation at fourth level pattern. The origin of the third level adjacent relation, with its organizational aspects of directional alignment and positional orientation, is the consequence of third level organizational initiation as it transforms into transformation point initiation when the transformation point takes on the role of adjacent place. The origin of the fourth level adjacent relation, and its organizational aspects, is the consequence of fourth level organizational initiation as it transforms into fourth level transformation point initiation when the stationary unit takes on an adjacent role. Because both cases of initiation are intrinsic existential-pathway-developments, the origins of the organizational aspects are all cases of self-organization (Figure 6.3).

Contact Relation

The second emergent stage is contact relation. When motion brings the units together, and the role of the transformation point develops the organizational initiation into transformation point initiation, fourth level adjacent rela-

Chapter 6: The Origin of Self-Organization

Contact

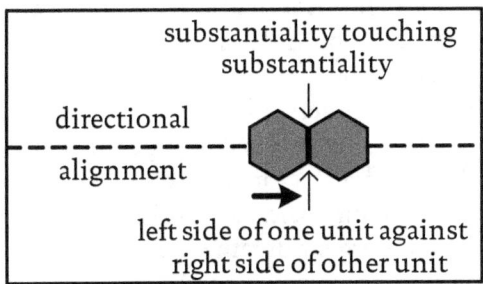

Figure 6.4 *As a supra-organizational relation, contact, the second emergent stage through the collision situation, is not a consequence of self-organization, but is a case of emergence alone. Nonetheless, the emergent organizational aspects of contact, such as the alignment of the relation and the positional orientations of the units, are products of self-organization.*

tion emerges. At fourth level, as soon as there is adjacent relation, there is contact relation. As a supra-organizational relation, contact is a consequence of emergence only, not of self-organization.

It is, nonetheless, a relation between two units and has intrinsic factors of organization. This relation occupies a spatial adjacent relation, constitutes a material adjacent relation, is existentially-dependent on the factor, adjacent relation, and has the same organizational aspects. There is a directional alignment relation and a positional orientation relation that occur with the contact relation (Figure 6.4). These are not the cases of directional alignment and positional orientation that occur with the adjacent relation between the occupied spatial places. Nor are they simply the organizational aspect of fourth level adjacent relation.

The contact relation does not exist without an additional role of substantiality—nor do its intrinsic organizational aspects. With substantiality something is-there that is not immaterial. In adjacent relation, it touches. Directional alignment and positional orientation develop from their roles with the adjacent relations of first and fourth level patterns of organization that occur foundationally in this situation to their roles based on substantiality in the contact relation. The locational aspect is the same for these three stages of development of directional alignment and positional orientation.

The development-of-origin of contact is organizational initiation, but in the developed form of adjacent-substantiality-primal-factor-initiation. This form of initiation is strictly intrinsic existential-pathway-development, and the emergent organizational aspects of contact are thus the result of self-organization.

Blocking

The next development of self-organization in the collision situation involves another development in the role of the nonimmateriality of substantiality—that something is-there, in space. Previous roles of material there-ness were (1) occupies spatial place, (2) materially differentiates pattern, (3) that which moves, and (4) contact. The role here is blocking. There is something there, in the way of the moving unit (Figure 6.5).

Blocking is a developed form of organizational initiation that is quite distinct from the prior form, which re-

Blocking

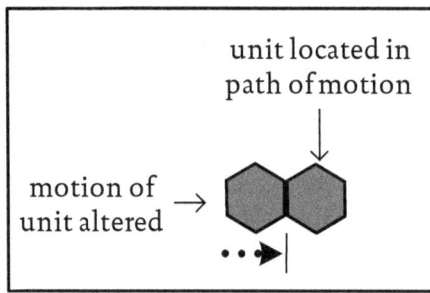

Figure 6.5 *Blocking is based on an organizational factor, location, augmented by substantiality. It is the first form of organizational initiation wherein a factor extrinsic to the initiator alters a factor intrinsic to the nature of the initiation—the first case of a consequence of self-organization occurring within the initiator itself.*

sults in contact, and also from the following form, which results in push. Prior forms of organizational initiation play their roles in blocking. It is organizational initiation because the situation involves a unit moving in relation to extrinsic factors, spatial place and another unit. It is a developed form of organizational initiation, transformation point initiation, because it involves a particular spatial place, which is occupied by another unit, where there occur abrupt changes in the existential-pathway-development of the situation. Up through the emergence of fourth level adjacent relation there is no direct interrelation between the two units. However, as soon as there is adjacent relation there is contact. It is again a developed form of organizational initiation, again by way of transformation point initiation, but this time with direct interrelation between the units. Contact originates through adjacent-substantiality-primal-factor-initiation.

The blocking situation has two material primary components, the units. Each has its own intrinsic existential-pathway-development. These two developmental pathways combined constitute the development of the situation. At first, from simple fourth level organizational relations through fourth level adjacent relation, these pathways are not directly combined. With the origin of contact, with its direct interrelation of the units, the pathways become directly interrelated.

To the stage of the emergence of fourth level adjacent relation, substantiality has played the roles of occupying space, differentiating pattern, and moving. But these roles of substantiality have not been the same in both existential-pathway-developments. Each unit plays an essentially equivalent role with the occupation of space and differentiation of pattern. That is not so with motion. Only one unit is moving, with significant consequences for the development of the situation.

Change existential-pathway-development is derived from continuing-existence and motion. Just prior to contact, the moving unit has both components of existential-pathway-development in its role with fourth level initiation. The stationary unit has only the one. This difference does not make a difference with contact relation. Contact is based simply on two units of substantiality in adjacent relation. It does make a difference with this next development of initiation.

In the blocking situation, the roles of the units and their substantiality are again not the same, and this difference is based on the difference in the origins of their existential-pathway-developments. In a collision, each unit alters the existential-pathway-development of the other. But each does so in a different manner, by way of a different form of initiation. In one case it is the role of substantiality in transformation point initiation without a role for motion. In the other case it is the role of substantiality in transformation point initiation in concert with motion.

Developmentally, blocking is prior to push. There can be no push without there being something there to push against. Push is existentially-dependent on blocking. However, comparing the two helps clarify the nature of blocking.

Change development goes with continuing-existence and motion. The continuing-existence of one unit cannot influence the continuing-existence of the other. The nature of continuing-existence cannot change. The noncoexistent-sequential-difference of the continuing-existence of anything that exists is always simultaneous and isomorphic with that of everything else that exists. Collision does not alter the foundational continuing-existence of any existing component of the situation.

It is motion that gets altered. Motion can go at various speeds and in various directions, and it is the quantitative and/or organizational aspects of speed and direction that get changed by blocking. This is a transformation point where the manner of change of changing extensional relations is altered. At this stage, though, they are altered by way of the role of the substantiality of the stationary unit in changing the motion of the other unit.

Chapter 6: The Origin of Self-Organization

The moving unit, because of its motion, pushes against the blocking unit. There is a development from the moving unit to the stationary unit. The motion of the one unit gives an aspect of unidirectionality to the existential-pathway-development of the collision situation. The blocking unit, lacking motion, does not push against the moving unit. There is no development from the blocking unit to the moving unit that is carried, transferred, or mediated by motion. Nothing goes from the blocking unit to the moving unit upon contact. (Not, at least, at this stage of analysis.) The stationary unit is just there, in the way. In this initiation situation, the initiator is the motion of the other unit. The role of the blocking unit is strictly that of a factor extrinsic to the initiator, simply an organizational role augmented by substantiality. Its role is passive, noncausal.

The stationary unit has an organizational role augmented by substantiality. This is the same as in its other roles. Because of its substantiality, the unit is-there such that it can occupy space. The nonimmateriality of its substantiality means it can differentiate a spatial place as a part of a pattern. Because there is something there, a discontinuous unit of substantiality, there is something there that can pass through space. It is substantiality that makes contact possible. And in the collision situation, it is the substantiality of the stationary unit that results in blocking.

This initiation situation is a moving unit in contact with a stationary unit located directly in the path of the motion. Because something is there, the moving unit cannot go on in the same manner as through immaterial space, which is incapable of influencing the motion. The motion is blocked.

The organizational role of the stationary unit, augmented by its substantiality, alters the motion of the other unit. It alters the role of the motion of the other unit in the existential-pathway-development of that unit, and by doing so, it changes the existential-pathway-development of the situation. Blocking alters the role of motion in existential-pathway-development. At this stage of the development of initiation situation, a factor extrinsic to the initiator changes the character of the initiator. This is the development-of-origin of a factor critical to the nature of complex self-organization—that matter reorganizes motion.

All the organizational aspects of this ongoing situation development play their roles in the origin of blocking and its emergent organization. The location and alignment of the motion, the location of the stationary unit, the alignments of the various adjacent relations and that of the contact relation, interrelate such that the event of blocking has an emergent intrinsic factor of alignment. These devel-

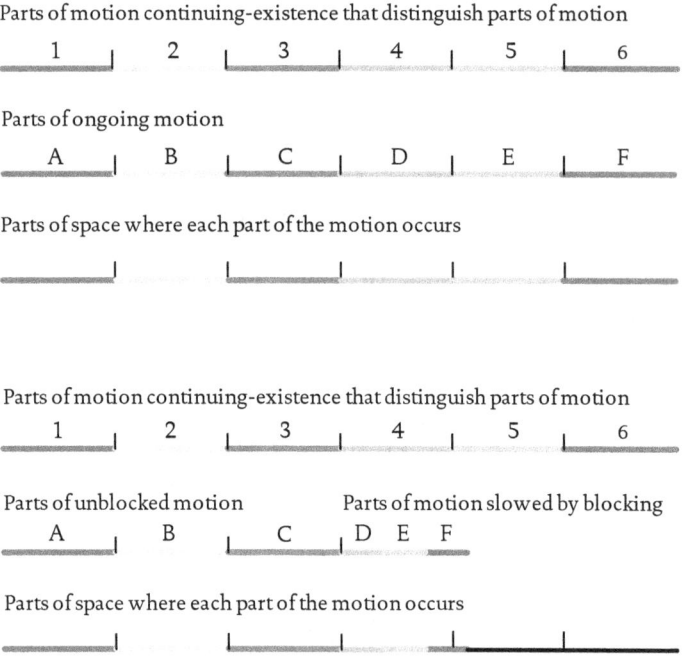

Figure 6.6 *Changed motion of the moving unit. When a moving unit is blocked and its speed reduced, the following parts of that motion will occur at different locations in space from what would have been the case without the blocking—a self-organized consequence of collision situation.*

opmentally prior factors are intrinsic to the collision situation, making the organizational factor of blocking alignment a consequence of self-organization.

The Changed Motion of the Moving Unit

In the blocking situation the motion is lessened, slowed or stopped, and thereby the organizational aspects of motion noncoexistent-sequential-difference are changed. Two of these organizational factors are (a) the size of the parts of motion and (b) the distance between parts. Motion is itself continuous with no intrinsically differentiated parts. There is, however, a feature of the existence of motion which does differentiate one part from another, which intrinsically makes one part of motion distinct from another—motion continuing-existence. New part of motion occurs simultaneously with new part of motion continuing-existence. During any particular part of the continuing-existence of a motion, there occurs a particular part of the ongoing motion itself. Motion continuing-existence is simultaneous and isomorphic to spatial continuing-existence. During any specific part or period of time, there occurs a specific part of an ongoing motion (Figure 6.6).

Continuing-existence, because it is simply continuance-of-being, is constant and cannot have any form of variance whatsoever. It cannot play a role in the variations, from part to part, of an ongoing existential-pathway-development. However, continuing-existence makes it possible for there to be other forms of change. Motion is different. It can have variance in that it can occur at different speeds. When a unit moves at a specific speed for a specific period of time, it traverses a specific distance in space. When that unit moves at a greater or slower speed for that same amount of time, it traverses a longer or shorter distance in space. The part of the motion that occurs with that specific part of time, and thus with a specific part of its own continuing-existence, can be longer or shorter, larger or smaller in size, depending on the speed of the unit.

When a stationary unit blocks the motion of another unit, the motion of that unit becomes less, and the parts of the motion, per part of its continuing-existence, are smaller. This is a change in the organization of the noncoexistent-sequential-difference of the motion. Because it is the consequence of factors intrinsic to the collision situation, it is a case of self-organization.

Shorter parts of motion are only one of several self-organized aspects that occur with this stage of collision. Another one that occurs as a consequence of blocking, again an aspect of motion itself, is how far along the path of the motion nonadjacent parts occur. When the speed is lessened and the parts are shorter, nonadjacent parts happen closer together.

There are also consequences extrinsic to the initiator that are results of self-organization. These, too, are changes in organizational aspects of noncoexistent-sequential-difference, changes involving extensional relations. For example, the rate of decrease of the distance relation between the moving unit and spatial places ahead on its path becomes slower. It takes longer, a greater amount of the continuing-existence of the motion, for it to reach those places. Another example is the changing rate of change of the direction relation between the moving unit and any spatial place it is passing. That rate continuously increases as the unit approaches the transformation point. With an unoccupied transformation point, the rate decreases as the unit leaves the area. With an occupied transformation point, the speed gets abruptly lessened, and in the case where the unit continues on past the transformation point, the rate of decrease in the amount the direction relation is changing will occur at a slower rate.

The location in space of a motion is part of the organizational relation of that motion with space. Since continuing-existence is constant, and the distance between nonadjacent parts of a motion that go with specific parts of its continuing-existence are shorter or longer depending on the speed of the motion, the parts of a motion take place in specific locations in space depending on the speed. When a moving unit is blocked and its speed reduced, the following parts of that motion will occur at different locations in space from what would have been the case without the blocking—a self-organized consequence of collision situation.

At the fourth level of pattern, at the origin of emergence, the motion factor of organizational initiation changes the organization of matter. It does so without the roles of contact, blocking, or causal push. This stage is characterized by motion changing the organization of matter simply by changing extensional relations. Within the collision situation, blocking changes the organization of motion. It does so with a role for contact, but without push. Motion changes the organization of matter, and blocking matter changes the organization of motion. This couplet of relations, here in precursor form, is the core of developed self-organization. Highly developed forms of self-organization, such as evolution, ontogeny, and rational thought, require the causal relation wherein motion, by way of contact and blocking, causes the reorganization of matter, and in the process is itself reorganized.

Push

To exist is to have organization. Emergent cause has emergent organization. Push is unidirectional. It is so, first, because it is a form of consequent-existence, second, because it is existentially-dependent on the noncoexistent-sequential-difference of continuing-existence, and third, because it is based, foundationally, on motion (Figure 6.7).

The organization of push in the collision situation is existentially-dependent on that of developmentally prior stages. In the case under consideration, the directional alignment of the relation and the positional orientation relation of push between the units are existentially-dependent on those relations as they occur in the contact relation, in the adjacent relation between the components, and in the underlying adjacent relations of the third and

Push

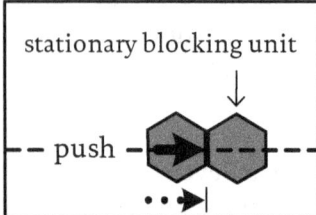

Figure 6.7 *While push itself is supra-organizational, it has emergent organizational factors due to self-organization. For example, unidirectionality and directional alignment that are existentially-dependent on their occurrence in motion, adjacent relation, contact, and blocking.*

first levels of pattern. The push relation occurs along the orientation of these relations. It occurs where they are. The units must occupy adjacent spatial places and thereby be adjacent themselves, and in contact. The push relation occurs across these relations, from one unit to the other, and does so in a unidirectional manner due to consequent-existence and motion. Extensional relations provide organization in space, the noncoexistent-sequential-difference of continuing-existence provides organization in time, and the moving substantiality of the one unit blocked by the stationary substantiality of the other unit provides organization of causal initiation, of push.

All these factors are intrinsic, from the places occupied, to the units with their substantiality, to the motion, to the alignments and orientation relations. Cause is emergent, and its unidirectional organization is a consequence of self-organization, in the form of collision-primal-factor-initiation.

The New Motion of the Blocking Unit

In space when a moving unit of matter pushes against a stationary unit, the stationary unit begins to move. The pressure from the one unit causes the other unit to move. The motion of this unit is emergent—a case of caused emergence.

The motion itself is supra-organizational, a consequence of emergence, not of self-organization. There are, however, organizational factors of this motion, just as there were for the motion of the unit that initiated the collision (Figure 6.8). Because the motion is emergent, its intrinsic organizational factors are emergent. At the development-of-origin of this motion, its factors are also cases of caused emergence.

Organizational factors of a motion are its direction in space, its speed, and the before and after relations of the parts of the motion's noncoexistent-sequential-difference. Speed plays an organizational role in relation to the before and after relations in that speed sets the size of a part per part of continuing-existence, and thus the distance between separate parts of the motion. Speed also plays an organizational role in relation to the direction of the motion in setting the current location of the ongoing motion in space, and thus the current extensional relations of the moving unit to spatial place and to anything else that exists in space.

New motion of the blocking unit

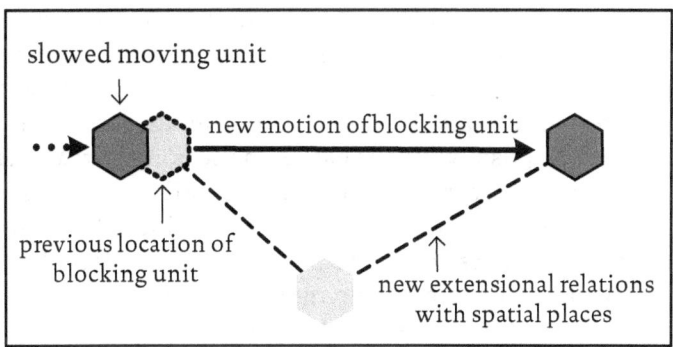

Figure 6.8 *The new motion of the blocking unit is supra-organizational, a consequence of emergence, not of self-organization. Factors of the new motion that are consequences of self-organization are direction, speed, sequential relations of the parts of the new motion, and new extensional relations with spatial place.*

The push gives the unit its motion, with the before and after relations of noncoexistent-sequential-difference intrinsic to the nature of motion. The other organizational qualities of the emergent motion, its speed, and in what direction it goes, are consequences of the nature of the developmentally prior moment of the situation's existential-pathway-development. They are consequences of the qualities of the moving unit and of its motion, and the qualities of the stationary unit and its lack of motion. These factors determine the nature of the push, and through the push, the specifics of the organizational factors of the new motion.

Some examples of the factors of the collision event that do or can play roles that determine consequent direction and speed are the existential quantity of each of the units, the speed of the moving unit and the direction of its motion, the shape and topography of each unit, the internal distribution of the density of the matter within them, their hardness and resiliency, and whether the collision is a direct hit or a glancing blow. Since these factors are intrinsic to the collision situation, the organizational aspects of the emergent motion are the consequence of self-organization. It is self-organization based on cause.

With its new motion, the unit gets new extensional relations with space. Its organizational relations with space are changed. It is also now in motion relative to the other unit, with changed fourth level extensional relations. The push, which is based on motion, reorganizes matter. At the origin of emergence, motion reorganizes matter. In the collision situation, the blocking role of substantiality changes the organization of the motion of the moving unit. Now here again motion, by way of push, reorganizes matter, a development of this factor.

With the collision situation, the early development of self-organization has attained a set of core qualities that more recognizably characterizes the developed forms. Motion reorganizes matter, and blocking matter reorganizes motion. Here contact plays a role in both stages and cause plays a role in the one stage. It is through the contact and causal relations that the qualities of the units and their motion, or lack of motion, determine the qualities of the consequent emergent factors, supra-organizational and organizational.

What goes before determines what follows, with consequent-existence, with continuing-existence, with motion, with emergence, with cause, and with self-organization.

Appendix 1
The Differences between the Extension of Spatial Place and the Continuing-Existence of Spatial Place

The mode-of being of the extensional aspect of spatial place is different from the mode-of-being of the continuing-existence of spatial place. The factors that play roles in the nature of space, (for example extension, voluminality, existential quantity, continuance, and sequentiality), are either entirely different or different in various respects from the factors that play roles in the continuing-existence of space. Even when the same factor plays a role in both cases, (for example quantity, continuance, or sequentiality), it does so in a different mode in each case.

Quantity, as it occurs with spatial-extension, is distinct, it has a different mode-of-being from the quantity of spatial continuance-of-being. All the quantity of infinite spatial extension exists simultaneously, coexistently. Continuance-of-being quantity occurs sequentially, noncoexistently. While the factors of extension and voluminality play roles in the mode-of-being of the existence of space, they do not do so in the mode-of-being of the continuing-existence of space, thus the quantity of spatial extension is voluminal, three-dimensional, while the quantity of continuance-of-being is nonextensional, nondimensional. The quantity of spatial extension does not change, neither increasing nor decreasing. The quantity of the continuing-existence of space is ever changing, ever increasing. The sequential-difference of extensional quantity is based on difference in location and does not involve change. The sequential-difference of continuance-of-being quantity is based on difference in existence, on noncoexistence, and is thus dependently based on change. While the factor quantity plays roles in both the existence of space and in the continuing-existence of space, the existential quantity of extension is distinct from the quantity of continuance-of-being.

The continuance that is an aspect of spatial extension is distinct, it has a different mode-of-being, from the continuance that is an aspect of the continuing-existence of spatial extension. The relation of the continuance to its associated quantity is different in the two modes-of-being.

In the spatial extension situation the quantity is primary and the continuance is secondary. The quantity of spatial extension is continuous. The continuous aspect is a characteristic of the quantity aspect. In the continuing-existence situation the continuance is primary and the quantity is secondary. That there is the continuance determines that there is the quantity of that continuance. The occurrence of the quantity is a consequence of the occurrence of the continuance.

In the spatial extension situation the quantity establishes the existential state of the continuance. All quantity of infinite spatial extension is coexistent. Thereby all continuance of that quantity is also coexistent. In the continuing-existence situation the continuance establishes the existential state of the quantity. Continuing-existence is noncoexistently sequential, and so is its quantity.

Because the quantity of spatial extension is coexistently continuous it constitutes an existing continuum. Adjacent to any continuous quantity of spatial extension there exists another part of the continuous quantity of space. With the continuance of extension there is always an aspect of existing continuum. Because the quantity of the continuing-existence of space is sequentially noncoexistent it does not constitute an existing continuum. During the current part of the continuing-existence of space, the adjacent parts of that continuance do not exist. With the continuance of the continuing-existence of space there is no aspect of existing continuum.

The parts of spatial extension have a different mode-of-being from the parts of the continuing-existence of spatial extension. The parts of spatial extension are coexistent, while the parts of the continuing-existence of space are noncoexistent. The parts of spatial existential quantity are parts of something that exists, something that is there, with extension and voluminality, while the parts of continuance-of-being quantity are the parts of the continuing-existence of that which exists with extension and voluminality. Because the parts of spatial extension are coexistent parts of an existing voluminal continuum of static place, they have specific distance and direction relations. These relations are geometric relations in that they can be summarized and analyzed by way of Euclidean geometry. Because the parts of spatial continuing-existence are not coexistent and are not parts of a continuum, nor of extensional place of any sort, they do not have distance and direction relations. There is no geometry to continuing-existence.

The sequentiality of spatial extension is distinct from the sequentiality of its continuing-existence. The sequential-difference of spatial extension is different from the sequential-difference of continuing-existence. The parts of spatial extension that constitute sequential-difference are coexistent, while the parts that constitute the sequential-difference of continuing-existence are not coexistent. The parts of spatial extension that constitute sequential-difference are existentially distinct because they are locationally distinct, while the parts that constitute the sequential-difference of continuing-existence are existentially distinct because they are noncoexistent. With coexistent-sequential-difference you can look at a whole sequence at one time, you can see the various parts simultaneously. With noncoexistent-sequential-difference you cannot look at a whole sequence at the same time, you cannot see the various parts simultaneously. Sequential-difference is geometrical in spatial extension, but is not so in continuing-existence.

The two forms of sequential-difference are also distinct in their relations to change and newness. With the sequential-difference of spatial extension there is no role for change or newness. The sequential-difference of spatial continuance-of-being, however, is a form of change. With the sequential-difference of continuing-existence there is continuing newness.

The organization of voluminal spatial extension has a different mode-of-being from the organization of its continuing-existence. The parts of extensional organization are coexistent, and their relations are those of coexistent components, for example the distance and direction relations are coexistent with the parts of space. The parts of the organization of continuance-of-being are sequentially noncoexistent, and their relations are those of sequentially noncoexistent components, for example the quantitatively specific after and before relations have no current existence. Spatial extension is three-dimensional, while continuing-existence is not dimensional at all. The organization of spatial extension is geometrical, while continuing-existence has no geometrical aspects to its mode-of-being. All the organizational aspects of spatial extension are static with no role for change of any sort, while continuing-existence is change with new part constantly coming into existence with new quantitatively specific after and before relations.

The extensional pattern of spatial place is distinct from the pattern of the noncoexistent-sequential-difference of its continuing-existence for the same reasons the organization of extension is distinct from the organization of continuing-existence.

Spatial place existential context is distinct from spatial continuance-of-being existential context. The mode-of-being of spatial place is appropriate as a context for the coexistent factors of the extension, but not for the noncoexistent factors of the continuing-existence, of that which exists in space. The mode-of-being of spatial continuance-of-being is appropriate as a context for the noncoexistent factors of other cases of continuing-existence, but not for the coexistent factors of extension. The context provided by coexistent spatial place alone is insufficient as a context for change, newness, motion, emergence, cause, and process. Noncoexistently occurring spatial continuance-of-being provides an adequate context for the change and newness that occurs with the continuing-existence of that which exists other than space, but it also does not do so for motion, emergence, cause, and process. The existential contexts provided by both spatial extension and spatial continuing-existence are required for there to be motion, emergence, cause, and process.

The relativity that occurs with spatial extension is distinct from the relativity that occurs with the continuing-existence of spatial extension. The relativity that occurs with spatial extension is between parts that are coexistent, have three-dimensional geometrical relations, are existentially distinct because they are locationally distinct, and are static without any roles in their existence or relative relations for change and newness. The relativity that occurs with continuing-existence is between parts that are not coexistent, that occur sequentially without dimensional geometrical relations, are existentially distinct because they are noncoexistent, and have roles for change and newness in their origin and in the origin of their relative relations.

Continuing-existence as a factor of reality is related to but distinct from existence itself as a factor of reality.

Appendix 2
Summary of Factors That Are Identical in Spatial and Material Continuing-Existence

The continuing-existence of space and that of matter are both are cases of existential initiation, existence initiating continuing-existence. In both cases it is continuance-of-being consequent-existence, the existence of what goes before determining, by way of continuing-existence, the existence of what follows. It is determinate consequent-existence in each case because it is the self-identity of what exists determining through its own continuing-existence the self-identity of what follows. With this second stage of initiation, the form is the same as with the spatial case. It is consequent-existence by way of continuing-existence, wherein that which exists continues to exist as itself, with no roles for other factors. With space and with matter it is a uniform, unidirectional continuance because it is continuance-of-being. With a primal-form existential initiator the initiation of continuing-existence is constant such that there occurs an ever-increasing quantity of the continuing-existence of space or of matter. Similar to the case with space, with matter something cannot come from nothing, and there is therefore something involved in the existence of matter that is eternal in that it has always existed, with the result that the increasing quantity of material continuing-existence had no beginning. In both cases the increasing quantity, as it occurs, is new, new sequentially noncoexistent part of ongoing continuing-existence, and that is change, the same form of change in both the spatial and material developmental pathways. A continuously ongoing uniform change with sequentially noncoexistent parts, a form of sequential-difference, will have quantitatively specific after and before relations between those parts, as is the case with the continuing-existence of both space and matter. The parts of the continuing-existence of space and of matter occur in noncoexistent sequence relative to one another, and thus the continuing-existence of both cases has the same organization. The continuing-existence of both space and matter are cases of pattern based on noncoexistent-sequential-difference, the parts of which have distinct self-identity differentiated by their individual existence and their noncoexistence, with those parts having organizational relations of sequentiality, unidirectionality, and specific after and before relations. A change development is a transition from one state of a situation to a following state in which continuance and noncoexistent-sequential-difference play roles and in which there is some form of enhancement which can be as simple as increased quantity of continuing-existence that has occurred, and this is the situation with the continuing-existence of space and of matter. In both the spatial case and the material case, there is a role for sequential enhancement, the occurrence of new part, and the material case is a development of this factor from the spatial case, an existential context dependent, nonpathway factor development.

Appendix 3
Two Lists of the Factors That Play Roles with Motion

1. Factors of Significance for Motion
Listed by type of factor.

A1. Extension [1 case, 1 form]:

> B1. Extension of spatial place.

A2. Continuance [5 cases, 3 forms]:

> B1. Coexistent continuance [1st of 3 forms];
>
> > C1. Coexistent continuance of spatial extension (1st of 5 cases).
>
> B2. Noncoexistent continuance based on continuing-existence [2nd of 3 forms];
>
> > C1. Noncoexistent continuance of spatial continuing-existence (2nd of 5 cases);
> >
> > C2. Noncoexistent continuance of material continuing-existence. (3rd of 5 cases);
> >
> > C3. Noncoexistent continuance of motion continuing-existence. (4th of 5 cases).
>
> B3. Noncoexistent continuance based on motion [3rd of 3 forms] (5th of 5 cases).

A3. Uniformity [5 cases, 3 forms]:

> B1. Uniform coexistent-sequential-difference [1st of 3 forms];
>
> > C1. Uniformity of spatial extension. (1st of 5 cases).
>
> B2. Uniform noncoexistent-sequential-difference [2nd of 3 forms];
>
> > C1. Uniformity of spatial continuing-existence (2nd of 5 cases);
> >
> > C2. Uniformity of material continuing-existence (3rd of 5 cases);

> C3. Uniformity of motion continuing-existence (4th of 5 cases).

B3. Uniformity of undisturbed motion [3rd of 3 forms] (5th of 5 cases).

A4. Parts [5 cases, 3 forms]:

> B1. Coexistent parts [1st of 3 forms];
>
>> C1. Parts of spatial extension (1st of 5 cases).
>
> B2. Noncoexistent parts of continuing-existence [2nd of 3 forms];
>
>> C1. Parts of spatial continuing-existence (2nd of 5 cases);
>>
>> C2. Parts of material continuing-existence (3rd of 5 cases);
>>
>> C3. Parts of motion continuing-existence (4th of 5 cases).
>
> B3. Noncoexistent parts of motion [3rd of 3 forms] (5th of 5 cases).

A5. Static and unchanging [1 case, 1 form]:

> B1. Static and unchanging character of spatial extension.

A6. Distinct individual self-identity [5 cases, 3 forms]:

> B1. Self-identity that is coexistent and an aspect of extension. [1st of 3 forms];
>
>> C1. Self-identity of parts of spatial extension. (1st of 5 cases).
>
> B2. Self-identity that is noncoexistent and an aspect of continuing-existence [2nd of 3 forms];
>
>> C1. Self-identity of parts of spatial continuing-existence (2nd of 5 cases);
>>
>> C2. Self-identity of parts of material continuing-existence (3rd of 5 cases);
>>
>> C3. Self-identity of parts of motion continuing-existence (4th of 5 cases).

> B3. Self-identity that is noncoexistent and an aspect of motion [3rd of 3 forms];
>> C1. Self-identity of parts of motion (5th of 5 cases).

A7. Coexistence [4 cases, 3 forms]:
> B1. Coexistent parts of a whole [1st of 3 forms];
>> C1. Coexistent parts of spatial extension (1st of 4 cases).
>
> B2. Coexistence relation between distinct factors [2nd of 3 forms];
>> C1. Coexistence relation between a unit of matter and motion (2nd of 4 cases).
>
> B3. Coexistence that is an aspect of the spatial existential-context relation [3rd of 3 forms];
>> C1. Coexistence of space and a unit of matter (3rd of 4 cases);
>>
>> C2. Coexistence of space and motion (4th of 4 cases).

A8. Sequentiality [5 cases, 3 forms]:
> B1. Sequentiality of extension [1st of 3 forms];
>> C1. Coexistent-sequential-difference of spatial extension (1st of 5 cases).
>
> B2. Sequentiality of continuing-existence [2nd of 3 forms];
>> C1. Noncoexistent-sequential-difference of spatial continuing-existence (2nd of 5 cases);
>>
>> C2. Noncoexistent-sequential-difference of material continuing-existence (3rd of 5 cases);
>>
>> C3. Noncoexistent-sequential-difference of motion continuing-existence (4th of 5 cases).
>
> B3. Sequentiality of motion [3rd of 3 forms];
>> C1. Noncoexistent-sequential-difference of motion (5th of 5 cases).

A9. Initiation. [4 cases, 2 forms]:

> B1. Initiator—Space;
>
>> C1. Initiates spatial continuing-existence (1st of 4 cases);
>>
>>> D1. Existential initiation [1st form].
>
> B2. Initiator—Matter;
>
>> C1. Initiates material continuing-existence (2nd of 4 cases);
>>
>>> D1. Existential initiation [1st form].
>
> B3. Initiator—Motion;
>
>> C1. Initiates motion continuing-existence (3rd of 4 cases);
>>
>>> D1. Existential initiation [1st form].
>>
>> C2. Initiates ongoing motion (4th of 4 cases);
>>
>>> D1. Primal factor Initiation [2nd form].

A10. Continuing-existence [3 cases of 1 form]:

> B1. Continuing-existence of space (1st of 3 cases);
>
> B2. Continuing-existence of matter (2nd of 3 cases);
>
> B3. Continuing-existence of motion (3rd of 3 cases).

A11. Noncoexistence (Noncoexistent forms of continuous sequential-difference) [4 cases of 2 forms]:

> B1. Noncoexistence based on continuing-existence [1st of 2 forms];
>
>> C1. Noncoexistent parts of spatial continuing-existence (1st of 4 cases);
>>
>> C2. Noncoexistent parts of material continuing-existence (2nd of 4 cases);
>>
>> C3. Noncoexistent parts of motion continuing-existence (3rd of 4 cases).
>
> B2. Noncoexistence based on motion [2nd of 2 forms];

C1. Noncoexistent parts of ongoing motion (4th of 4 cases).

A12. Change [4 cases, 2 forms]:

> B1. Change of continuing-existence [1st of 2 forms];
>
>> C1. Change of spatial continuing-existence (1st of 4 cases);
>>
>> C2. Change of material continuing-existence (2nd of 4 cases);
>>
>> C3. Change of motion continuing-existence (3rd of 4 cases).
>
> B2. Change that is motion itself [2nd of 2 forms] (4th of 4 cases).

A13. Change in aspect of self-identity [4 cases, 2 forms]:

> B1. Change in self-identity of continuing-existence [1st of 2 forms];
>
>> C1. Change in self-identity of spatial continuing-existence (1st of 4 cases);
>>
>> C2. Change in self-identity of material continuing-existence (2nd of 4 cases);
>>
>> C3. Change in self-identity of motion continuing-existence (3rd of 4 cases).
>
> B2. Change in self-identity of motion [2nd of 2 forms] (4th of 4 cases).

A14. Unidirectionality of change [4 cases, 2 forms]:

> B1. Unidirectionality of continuing-existence [1st of 2 forms];
>
>> C1. Unidirectionality of spatial continuing-existence (1st of 4 cases);
>>
>> C2. Unidirectionality of material continuing-existence (2nd of 4 cases);
>>
>> C3. Unidirectionality of motion continuing-existence (3rd of 4 cases).

B2. Unidirectionality of motion [2nd of 2 forms] (4th of 4 cases).

A15. Newness [4 cases, 2 forms]:

> B1. Newness that occurs with continuing-existence [1st of 2 forms];
>
>> C1. New part of spatial continuing-existence (1st of 4 cases);
>>
>> C2. New part of material continuing-existence (2nd of 4 cases);
>>
>> C3. New part of motion continuing-existence (3rd of 4 cases).
>
> B2. Newness that occurs with motion itself [2nd of 2 forms];
>
>> C1. New part of motion (4th of 4 cases).

A16. Simultaneity [6 cases, 2 forms]:

> B1. Simultaneity of cases of continuing-existence [1st of 2 forms];
>
>> C1. Simultaneity of the continuing-existence of the coexistent parts of space (1st of 6 cases);
>>
>> C2. Simultaneity of the continuing-existence of matter with the continuing-existence of space (2nd of 6 cases);
>>
>> C3. Simultaneity of the continuing-existence of motion with the continuing-existence of matter (3rd of 6 cases);
>>
>> C4. Simultaneity of the continuing-existence of motion with the continuing-existence of space (4th of 6 cases).
>
> B2. Simultaneity of new part of motion with new part of continuing-existence [2nd of 2 forms];
>
>> C1. Simultaneity of new part of ongoing motion with new part of material continuing-existence (5th of 6 cases);

C2. Simultaneity of new part of ongoing motion with new part of spatial continuing-existence (6th of 6 cases).

2. Developmental List of the Factors That Play Roles with Motion

Listed by mode-of-being (space, matter, motion), and then by extensional, continuing-existence, and coexistence roles, with the cases of each type of factor numbered.

The number sequences of the factors are not exactly the same between the two lists because of differences in the overall organization of the lists.

A1. Space:

    B1. Spatial extensional factors;

        C1. Spatial extension (1 of 1) is continuous (1st of 5);

        C2. Continuous spatial extension is uniform (1st of 5);

        C3. Continuous spatial extension has parts (1st of 5);

        C4. The parts of spatial extension have distinct individual self-identity (1st of 5);

        C5. The parts of spatial extension are coexistent (1st of 4);

        C6. The distinct coexistent parts of spatial extension are organized sequentially (1st of 5) with one another, constituting coexistent-sequential-difference (1 of 1);

        C7. The parts of spatial extension are static and unchanging (1 of 1).

    B2. Spatial continuing-existence factors;

        C1. The existence of space initiates (1st of 4) spatial continuing-existence (1st of 3);

        C2. Spatial continuing-existence is continuous (2nd of 5);

C3. Spatial continuing-existence is uniform (2nd of 5);

C4. Spatial continuing-existence has parts (2nd of 5);

C5. The parts of spatial continuing-existence have distinct individual self-identity (2nd of 5);

C6. The parts of spatial continuing-existence are noncoexistent (1st of 4);

C7. The distinct noncoexistent parts of spatial continuing-existence are organized sequentially (2nd of 5) with one another, constituting noncoexistent-sequential-difference (1st of 4);

C8. Spatial continuing-existence is a form of change (1st of 4);

C9. Spatial continuing-existence has change in self-identity (1st of 4);

C10. The change that is spatial continuing-existence is unidirectional (1st of 4);

C11. The unidirectional change of spatial continuing-existence results in continuously new (1st of 4) part of spatial continuing-existence;

C12. The continuing-existence of the parts of space is simultaneous (1st of 6).

A2. Matter:

B1. Material extensional factors;

C1. Because a unit of matter can play its role in motion as a whole, the internal extensional factors of that unit can at first be ignored in the analysis of the role of matter in the process of motion.

B2. Material continuing-existence factors;

C1. The existence of matter initiates (2nd of 4) material continuing-existence (2nd of 3).

C2. Material continuing-existence is continuous (3rd of 5);

C3. Material continuing-existence is uniform (3rd of 5);

C4. Material continuing-existence has parts (3rd of 5);

C5. The parts of material continuing-existence have distinct individual self-identity (3rd of 5);

C6. The parts of material continuing-existence are noncoexistent (2nd of 4);

C7. The distinct noncoexistent parts of material continuing-existence are organized sequentially (3rd of 5) with one another, constituting noncoexistent-sequential-difference (2nd of 4);

C8. Material continuing-existence is a form of change (2nd of 4);

C9. Material continuing-existence has change in self-identity (2nd of 4);

C10. The change that is material continuing-existence is unidirectional (2nd of 4);

C11. The unidirectional change of material continuing-existence results in continuously new (2nd of 4) part of material continuing-existence.

B3. Coexistence of matter with space;

C1. Matter is coexistent (2nd of 4) with space;

C2. The continuing-existence of matter is simultaneous with the continuing-existence of space (2nd of 6).

A3. Motion:

B1. Motion extensional factors;

C1. The existence of motion initiates (3rd of 4) ongoing motion (First of three types of consequents);

C2. Motion is continuous (4th of 5);

C3. Undisturbed motion is uniform (4th of 5);

C4. Motion has parts (4th of 5);

C5. The parts of motion have distinct individual self-identity (4th of 5);

C6. The parts of motion are noncoexistent (3rd of 4);

C7. The distinct noncoexistent parts of motion are organized sequentially (4th of 5) with one another, constituting noncoexistent-sequential-difference (3 of 4);

C8. Motion is a form of change (3rd of 4);

C9. Motion has change in self-identity (3rd of 4);

C10. The change that is motion is unidirectional (3rd of 4);

C11. The unidirectional change of motion results in continuously new (3rd of 4) part of motion.

B2. Motion continuing-existence factors;

C1. The existence of motion initiates (4th of 4) motion continuing-existence (3rd of 3) (Second of three types of consequents);

C2. Motion continuing-existence is continuous (5th of 5);

C3. Motion continuing-existence is uniform (5th of 5);

C4. Motion continuing-existence has parts (5th of 5);

C5. The parts of motion continuing-existence have distinct individual self-identity (5th of 5);

C6. The parts of motion continuing-existence are noncoexistent (4th of 4);

C7. The distinct noncoexistent parts of motion continuing-existence are organized sequentially (5th of 5) with one another, constituting noncoexistent-sequential-difference (4th of 4);

C8. Motion continuing-existence is a form of change (4th of 4);

C9. Motion continuing-existence has change in self-identity (4th of 4);

C10. The change that is motion continuing-existence is unidirectional (4th of 4);

C11. The unidirectional change of motion continuing-existence results in continuously new (4th of 4) part of motion continuing-existence.

B3. Coexistence of motion with matter and with space;

C1. Coexistence (3rd of 4) relation between a unit of matter and motion;

D1. New part of ongoing motion is simultaneous (3rd of 6) with new part of material continuing-existence;

D2. The continuing-existence of motion is simultaneous (4th of 6) with the continuing-existence of matter.

C2. Coexistence (4th of 4) of space and motion.

D1. New part of ongoing motion is simultaneous (5th of 6) with new part of spatial continuing-existence;

D2. The continuing-existence of motion is simultaneous (6th of 6) with the continuing-existence of space.

Appendix 4
Summaries of the Development of Some of the Individual Factors that Occur in the Foundational Development of Reality

Contents of Appendix 4

Introduction
1. Change
2. Change in Aspect of Self-Identity
3. Coexistence
4. Combinatorial Enhancement
5. Consequent-Existence
6. Continuity
7. Determinate-Reality
8. Initiation
9. Pattern of Extensional Factors
10. Simultaneity
11. Unidirectionality
12. Uniformity

Introduction

At the known foundations of reality there are several universal and near universal stages of development, and most of the factors that exist and play roles there develop through most of these stages. Hence, these stages show up repeatedly in the various summaries of the development of the individual factors. These stages are:

1. The existence of space;
2. The continuing-existence of space;
3. The existence of matter;
4. The continuing-existence of matter;

5. The existence of motion;
6. The continuing-existence of motion;
7. The occurrence of transformation points, and;
8. The occurrence of emergence.

The examples of factor development given here are not list mapped beyond the development-of-origin of emergence. Further stages described or mentioned in this book are:

9. Contact;
10. Cause;
11. Open through-flow, and;
12. Coherent relation.

Change

First Form
Change, noncoexistent-sequential-difference, of continuing-existence—

1. Of space;
2. Of matter, and;
3. Of motion.

Second Form
Change, noncoexistent-sequential-difference, of motion.

Third Form
Change, noncoexistent-sequential-difference, of extensional relations between a moving unit of matter and spatial place—change of:

1. Occupation/location relation;
2. Direction relations;
3. Distance relations, and;
4. Positional orientation relations.

Appendix 4

Fourth Form
Change, noncoexistent-sequential-difference, of pattern of extensional relations between a moving unit of matter and spatial place.

Fifth Form
Change, noncoexistent-sequential-difference, of extensional relations between a spatial place and a moving unit that has reached a transformation point—

> First Case: Unit passing by the spatial place, and passing through the point of shortest distance relation.
>
> 1. Change in manner of change—
>
>   1. Of distance relation from decreasing to increasing;
>
>   2. Of direction relation from increasing rate of change to decreasing rate of change, and;
>
>   3. Of orientation relation from increasing rate of change to decreasing rate of change.
>
> Second Case: Unit moving directly at the spatial place and passing through it, thus passing through the point of least distance relation.
>
> 1. Change in manner of change—
>
>   1. Of distance relation from decreasing to increasing;
>
> 2. Noncoexistent-sequential-difference—
>
>   1. Of direction relation from the orientation prior to reaching the spatial place and the transformation point to the opposite orientation after passing through;

2. Of positional orientation relation from the orientation prior to reaching the spatial place and the transformation point to the opposite orientation after passing through, and;
3. Of the nonexistence of an occupation relation between the unit and the spatial place, to the existence of an occupation relation when the unit reaches the spatial place and the transformation point, to the nonexistence of that relation after the unit moves on.

Sixth Form

Change, noncoexistent-sequential-difference, of pattern of extensional relations between a spatial place and a moving unit that has reached a transformation point—

> First Case: Unit passing by the spatial place, and passing through the point of shortest distance relation.
>
> 1. Change in pattern of extensional relations through change in manner of change of the relational components of the pattern.
>
> Second Case: Unit moving directly at the spatial place and passing through it, thus passing through the point of least distance relation.
>
> 1. Change in pattern of extensional relations through change in manner of change, change in orientation, or change in existential status, of the relational components of the pattern.

Appendix 4

Seventh Form
Change, noncoexistent-sequential-difference, of extensional relations between a moving unit of matter and a stationary unit—change of:

1. Direction relations;
2. Distance relations, and;
3. Positional orientation relations.

Eighth Form
Change, noncoexistent-sequential-difference, of pattern of extensional relations between a moving unit and a stationary unit.

Change in Aspect of Self-Identity

First Form
Change in an aspect of self-identity of continuing-existence.

1. Change from individually distinct part to following noncoexistent individually distinct part of:
    1. Spatial continuing-existence;
    2. Material continuing-existence, and;
    3. Motion continuing-existence.

Second Form
Change in an aspect of self-identity of motion.

1. Change from individually distinct part to following noncoexistent individually distinct part of motion.

Third Form
Change in an aspect of self-identity of a situation—the extensional relations between a moving unit of matter and spatial place.

1. Change from an individually distinct extensional relation to the immediately following, noncoexistent, individually distinct extensional relation—change from:
    1. One occupation/location relation to another;
    2. One direction relation to another;
    3. One distance relation to another, and;
    4. One positional orientation relation to another.

Fourth Form
Change in an aspect of self-identity of a situation—the pattern of extensional relations between a moving unit of matter and spatial place.

1. Change from an individually distinct pattern of extensional relations to the immediately following, noncoexistent, individually distinct pattern of extensional relations.

Fifth Form
Change in an aspect of self-identity of a situation—the extensional relations between a spatial place and a moving unit of matter which has arrived at a transformation point.

First Case: Unit passing by the spatial place, and passing through the point of shortest distance relation.

Transformation, change in manner of change, of extensional relations between the moving unit and the spatial place—transformation of:

1. Distance relation from decreasing to increasing;
2. Direction relation from increasing rate of change to decreasing rate of change, and;
3. Orientation relation from increasing rate of change to decreasing rate of change.

Second Case: Unit moving directly at the spatial place and passing through it, thus passing through the point of least distance relation.

Transformation, change in manner of change, in orientation, or in existence, of extensional relations between the moving unit and the spatial place—transformation of:

1. Distance relation from decreasing, to nonexistent, to increasing;
2. Direction relation, reversal of orientation;
3. Positional orientation relation, reversal of orientation from leading side to trailing side, and;
4. Occupation relation from nonexistent, to existing, to nonexistent.

Sixth Form

Change in an aspect of self-identity of a situation—the pattern of extensional relations between a spatial place and a moving unit which has arrived at a transformation point.

First Case: Unit passing by the spatial place, and passing through the point of shortest distance relation.

1. Transformation, change in manner of change, of pattern of extensional relations between a moving unit and a spatial place.

Second Case: Unit moving directly at the spatial place and passing through it, thus passing through the point of least distance relation.

1. Transformation of pattern of extensional relations between the moving unit and the spatial place, involving change in manner of change, in orientation, or in existence, of extensional relations.

Seventh Form

Change in an aspect of self-identity of a situation—the extensional relations between a moving unit and a stationary unit.

Change from an individually distinct extensional relation between the two units to the immediately following, noncoexistent, individually distinct extensional relation—change from:

1. One direction relation to another;
2. One distance relation to another, and;
3. One positional orientation relation to another.

Eighth Form

Change in an aspect of self-identity of a situation—the pattern of extensional relations between a moving unit and a stationary unit.

1. Change from an individually distinct pattern of extensional relations between the two units to the immediately following, noncoexistent,

individually distinct pattern of extensional relations.

Coexistence

Coexistence of infinite spatial place.

> Coexistence of the parts of space:
> 1. Of adjacent spatial places, and;
> 2. Of nonadjacent spatial places.

Coexistence of the matter of a primal unit.

> Coexistence of the parts of a primal unit of matter:
> 1. Of adjacent parts of a unit, and;
> 2. Of nonadjacent parts of a unit.

Coexistence of space and one static unit of matter.

Coexistence relation between a unit of matter and motion.

Coexistence of space and motion.

Coexistence of space, matter, and motion.

Coexistence of space and two static units of matter.

Coexistence of space, one static unit of matter, and one moving unit of matter.

Combinatorial Enhancement

(The coexistence of the parts of space.)

The combinatorial enhancement is:
1. Relative existence.

(Coexistence of adjacent parts of space.)

The combinatorial enhancements are:
1. Adjacent relation;
2. Direction relations between adjacent parts of space, (but no distance relations), and;
3. Locationally differentiated pattern of adjacent immaterial spatial places, with their direction relations.

(Coexistence of nonadjacent parts of space.)

The combinatorial enhancements are:
1. Nonadjacent relation;
2. Direction relations between nonadjacent parts of space;
3. Distance relations between nonadjacent parts of space, and;
4. Locationally differentiated pattern of nonadjacent immaterial spatial places, with their direction and distance relations.

(The coexistence of the parts of a primal unit of matter.)

The combinatorial enhancement is:
1. Relative existence.

(Coexistence of adjacent parts of a unit.)

The combinatorial enhancements are:

1. Adjacent relation;
2. Contact relation between adjacent parts of continuous substantial existential quantity of a primal unit;
3. Direction relations between adjacent parts of a unit, (but no distance relations), and;
4. Locationally differentiated pattern of adjacent parts of a unit, with their direction relations.

(Coexistence of nonadjacent parts of a unit.)

The combinatorial enhancements are:

1. Nonadjacent relation;
2. Direction relations between nonadjacent parts of a unit;
3. Distance relations between nonadjacent parts of a unit, and;
4. Locationally differentiated pattern of nonadjacent parts of a primal unit, with their direction and distance relations.

(Coexistence of space and one static unit of matter.)

The combinatorial enhancement is:

1. Existential context relation, wherein all qualities of matter occur within and conform to the qualities of space.

(Coexistence of the extension of primal matter with the extension of space.)

The combinatorial enhancements are:

1. The occupation relation between a unit of matter and the spatial place where it exists;

> 2. The direction relations of that unit to the rest of space;
> 3. The distance relations of that unit to the rest of space;
> 4. The positional orientation relation of the unit to static space, and;
> 5. In part materially and in part immaterially differentiated pattern of the unit with the parts of the rest of space.

(Coexistence relation between a unit of matter and motion.)

> (Coexistence of the matter with continuously new different part of the motion.)
>
>> (While there are apparently no secondary coexistence relations, which can be considered as combinatorial enhancements intrinsic to this coexistence relation, there is the directionally oriented existential-dependency relation of the motion on the unit of matter whose role is required for there to be motion.)

(Coexistence of space and motion.)

> (New combining of a motion with a continuous sequence of spatial location.)
>
>> The combinatorial enhancement is:
>>
>>> New location relation of the motion to space.

(Coexistence of space, a unit of matter, and motion.)

> (New combining of a unit of matter with spatial location as a consequence of motion.)

The combinatorial enhancement is:

1. New occupation relation.

(Coexistence of two static units of matter.)

The combinatorial enhancement is:

1. Relative existence of individual units of matter.

(Coexistence of adjacent units.)

The combinatorial enhancements are:

1. Adjacent relation;
2. Contact relation between adjacent units;
3. Direction relations between adjacent units, (but no distance relations);
4. Positional orientation relation between adjacent units;
    a. Direction relations between the parts of one unit and the parts of an adjacent unit;
    b. Distance relations between the parts of one unit and the parts of an adjacent unit.
5. Materially differentiated pattern of adjacent units, with their direction and positional orientation relations.

(Coexistence of nonadjacent units.)

The combinatorial enhancements are:

1. Nonadjacent relation;
2. Direction relations between nonadjacent units;

3. Distance relations between nonadjacent units;
4. Positional orientation relation between nonadjacent units;
    a. Direction relations between the parts of one unit and the parts of a nonadjacent unit;
    b. Distance relations between the parts of one unit and the parts of a nonadjacent unit.
5. Materially differentiated pattern of nonadjacent units, with their direction, distance, and positional orientation relations.

(Coexistence of space, one static unit of matter, and one moving unit of matter—The development-of-origin of emergence.)

The combinatorial enhancement is:

1. Emergence, the creative process whereby newly existing patterns of material organization come into existence.

(Emergence from combining, which at this stage is moving closer together.)

The combinatorial enhancement is:

1. Enhancement of group character of a situation of coexistent units—the coming into existence of the pattern of material organization, group.

Consequent-Existence

First Form
The existence of continuing-existence as a consequence of existence.

1. Spatial continuing-existence as a consequence of the existence of space;
2. Material continuing-existence as a consequence of the existence of matter;
3. Motion continuing-existence as the consequence of the existence of motion.

Second Form
The existence of ongoing motion as a consequence of the existence of motion.

Third Form
The existence of sequentially new extensional relations between a moving unit of matter and spatial place as a consequence of the motion of the unit in relation to space.

1. New occupation/location relations as a consequence of the motion in relation to the sequence of spatial place in the path of the moving unit;
2. New direction relations as a consequence of the motion in relation to the directions between the spatial places occupied by the moving unit and other spatial places;
3. New distance relations as a consequence of the motion in relation to the distances between the spatial places occupied by the moving unit and other spatial places, and;
4. New positional orientation relations as a consequence of the motion in relation to the positional orientation of the unit to spatial places.

Fourth Form
The existence of new pattern of extensional relations between a moving unit of matter and spatial place as a consequence of motion in relation to the direction, distance, and positional orientation relations of the unit to spatial place.

Fifth Form
Existence of abrupt transformations in extensional relations between a moving unit and a spatial place as a consequence of a transformation point.

> First Case: Unit passing by the spatial place, and passing through the point of shortest distance relation.
>
> 1. Abrupt change in manner of change of extensional relations as a consequence of the unit passing through the point of closest proximity to the spatial place—change of:
>   1. Distance relation, from decreasing to increasing;
>   2. Direction relation, from increasing rate of change to decreasing rate of change, and;
>   3. Positional orientation relation, from increasing rate of change to decreasing rate of change.
>
> Second Case: Unit moving directly at the spatial place and passing through it, thus passing through the point of least (zero) distance relation.
>
> 1. Abrupt change in manner of change of an extensional relation as a consequence of the unit passing through the spatial place and the point of zero distance relation—change of:
>   1. Distance relation, from decreasing to increasing.

2. Abrupt change in orientation of extensional relations as a consequence of the unit passing through the spatial place and the point of zero distance relation—change of:
    1. Direction relation, from the orientation prior to reaching the spatial place to the opposite orientation after passing through that place;
    2. Positional orientation relation, from that prior to passing through spatial place transformation point to the opposite, subsequent to passing through, and;
3. Abrupt change in existence status of an extensional relation as a consequence of the unit passing through the spatial place and the point of zero distance relation—change of:
    1. Occupation relation, from the nonexistence of an occupation relation between the unit and the spatial place, to the existence of an occupation relation when the unit reaches the spatial place, to the nonexistence of that relation after the unit moves on.

Sixth Form

Existence of abrupt transformations in pattern of extensional relations between a moving unit and a spatial place as a consequence of a transformation point.

> First Case: Unit passing by the spatial place, and passing through the point of shortest distance relation.
>
> 1. Abrupt change in manner of change of pattern of extensional relations, with

components transforming from decreasing to increasing and from increasing to decreasing, as a consequence of the unit passing through the point of closest proximity to the spatial place.

Second Case: Unit moving directly at the spatial place and passing through it, thus passing through the point of least (zero) distance relation.

1. Abrupt change in manner of change of pattern of extensional relations, with components transforming from decreasing to increasing, from one orientation to the opposite orientation, and from nonexistence to existence to nonexistence, as a consequence of the unit passing through the spatial place, with that spatial place playing the role of a transformation point.

Seventh Form

The existence of sequentially new extensional relations, between a moving unit and another unit that is not moving, as a consequence of the motion of the one unit relative to the other unit—development-of-origin for emergence.

1. New direction relations as a consequence of the motion in relation to the directions between the spatial places occupied by the moving unit and the spatial place occupied by the other unit, and thus in relation to the direction relations between the two units;

2. New distance relations as a consequence of the motion in relation to the distances between the spatial places occupied by the moving unit and the spatial place occupied by the other unit, and thus in relation to the distance relations between the two units, and;

3. New positional orientation relations as a consequence of the motion in relation to the positional orientation of the moving unit to the stationary unit.

Eighth Form
The existence of new pattern of extensional relations, between a moving unit of matter and a stationary unit, as a consequence of the motion of the one unit in relation to the direction, distance, and positional orientation relations between the two units. (Development-of-origin for emergence.)

Continuity

First Form
Continuous aspect of the extension of spatial place.
(Spatial place is a continuum, a form of continuance.)

Second Form
Continuous aspect of the extension of primal matter.
(Primal matter is a second form of continuance.)

Third Form
Continuous aspect of motion.
(Motion is a third form of continuance.)

Fourth Form
Continuous aspect of continuing-existence:

1. Of space;
2. Of matter, and;
3. Of motion.

(Continuing-existence is a fourth form of continuance.)

Sixth Form
Continuous aspect of changing extensional relations between a moving unit of matter and spatial place—continuous change of:

1. Occupation/location relation;
2. Direction relations;
3. Distance relations, and;
4. Positional orientation relations.

Seventh Form
Continuous aspect of changing pattern of extensional relations between a moving unit and spatial place.

Eighth Form
Continuous aspect of changing extensional relations between a moving unit of matter and a stationary unit—continuous change of:

1. Direction relations;
2. Distance relations, and;
3. Positional orientation relations.

Ninth Form
Continuous aspect of changing pattern of extensional relations between a moving unit and a stationary unit.

Determinate-Reality

First Form
Determinate consequent-existence by way of continuing to exist.

Determinate initiation of:
1. Spatial continuing-existence;
2. Material continuing-existence, and;
3. Motion continuing-existence.

Second Form
Determinate consequent-existence by of way motion continuing to be itself, a form of noncoexistent-sequential-difference, with what goes before determining what follows by continuing to be what it is.

Determinate initiation of ongoing motion.

Third Form
Determinate consequent-existence by way of motion in relation to the extensional relations between a moving unit and spatial place.

Determinate initiation of change of extensional relations between a moving unit of matter and spatial place—change of:
1. Occupation/location relations;
2. Direction relations;
3. Distance relations, and;
4. Positional orientation relations.

Fourth Form
Determinate consequent-existence by way of motion in relation to the pattern of extensional relations of the unit to spatial place.

Determinate initiation of change of pattern of extensional relations between a moving unit of matter and spatial place.

Fifth Form

Determinate consequent-existence by way of motion in relation to the extensional relations between a spatial place and a moving unit that passes through a transformation point.

> First Case: Unit passing by a spatial place, and passing through the point of shortest distance relation, the transformation point.
>
> 1. Determinate initiation of abrupt change in the manner of change of the extensional relations between the moving unit and the spatial place—determinate change of:
>   1. Distance relation, from decreasing to increasing;
>   2. Direction relation, from increasing rate of change to decreasing rate of change, and;
>   3. Positional orientation relation, from increasing rate of change to decreasing rate of change.
>
> Second Case: Unit moving directly at the spatial place and passing through it, thus passing through the point of least (zero) distance relation, the transformation point.
>
> 1. Determinate initiation of abrupt change in the manner of change of an extensional relation between the moving unit and the spatial place—determinate change of:
>   1. Distance relation, from decreasing to nonexistence, then reoccurrence and increasing.
> 2. Determinate initiation of abrupt change in orientation of extensional relations

between the moving unit and the spatial place—determinate change of:

1. Direction relation from existing, to not existing, to existing again but with opposite orientation, and;
2. Positional orientation relation from that prior to passing through the transformation point to the opposite, subsequent to passing through.

3. Determinate initiation of abrupt change in existence status of an extensional relation between the moving unit and the spatial place—determinate change of:

1. Occupation relation from not existing, to existing, to not existing again.

Sixth Form

Determinate consequent-existence by way of motion in relation to the pattern of extensional relations between a spatial place and a moving unit that passes through a transformation point.

First Case: Unit passing by a spatial place, and passing through the point of shortest distance relation, the transformation point.

Determinate initiation of abrupt transformation of uniform pattern of noncoexistent-sequential-difference of changing pattern of extensional relations to a different uniform pattern of noncoexistent-sequential-difference of changing pattern of extensional relations, which continues without further transformation.

Second Case: Unit moving directly at the spatial place and passing through it, thus passing through the point of least (zero) distance relation, the transformation point.

Determinate initiation of abrupt transformation of uniform pattern of noncoexistent-sequential-difference of changing pattern of extensional relations to a different uniform pattern of noncoexistent-sequential-difference of changing pattern of extensional relations, which continues without further transformation.

Seventh Form

Determinate consequent-existence by way of the motion of a unit of matter in relation to a stationary unit, in relation to the spatial location and positional orientation of the stationary unit, in relation to the extensional relations between the two units.

Determinate initiation of change of extensional relations between a moving unit of matter and a stationary unit—determinate change of:

1. Direction relations;
2. Distance relations, and;
3. Positional orientation relations.

Eighth Form

Determinate consequent-existence by way of motion in relation to the pattern of extensional relations of the moving unit to the static unit.

Determinate initiation of change of pattern of extensional relations between a moving unit and another unit that is not moving—development-of-origin for emergence.

Appendix 4

Initiation

Initiator—Space. (Universal foundational initiator, to which all other initiation conforms)
   1. Initiates continuing-existence of space. (Primal-form existential initiation—intrinsic consequence).

Initiator—Matter.
   1. Initiates continuing-existence of matter. (Primal-form existential initiation—intrinsic consequence).

Initiator—Motion.
   1. Initiates continuing-existence of motion. (Primal-factor existential initiation—intrinsic consequence).
   2. Initiates ongoing motion. (Primal factor initiation—intrinsic consequence).

Initiation situation—Motion of a unit of matter in relation to spatial place.
   1. Initiates change in extensional relations between a moving unit and spatial place. (Organizational initiation—extrinsic consequence)

      Change in:
      1. Occupation/location relations;
      2. Distance relations;
      3. Direction relations, and;
      4. Positional orientation relations.

   2. Initiates change in pattern of extensional relations between a moving unit and spatial place.

(Organizational initiation—extrinsic consequence).

Initiation situation—Motion of a unit of matter in relation to a particular spatial place and in relation to a transformation point. (Transformation point initiation).

1. Initiates abrupt transformations in extensional relations between the moving unit and the spatial place. (Organizational initiation—extrinsic consequence—existential-pathway-transformational-development).

First Case: Unit passing by a spatial place, and passing through the point of shortest distance relation, the transformation point.

1. Initiates abrupt changes in the manner of change of:
    1. Distance relation, from decreasing to increasing;
    2. Direction relation, from increasing rate of change to decreasing rate, and;
    3. Positional orientation relation, also from increasing rate of change to decreasing rate.

Second Case: Unit moving directly at the spatial place and passing through it, thus passing through the point of least (zero) distance relation, the transformation point.

1. Initiates abrupt change in the manner of change of:
    1. Distance relation, from decreasing to nonexistence, then reoccurrence and increasing.
2. Initiates abrupt change in orientation of:

1. Direction relation from existing, to not existing, to existing again but with opposite orientation;
2. Positional orientation relation from that prior to passing through the spatial place transformation point to the opposite, subsequent to passing through.
3. Initiates abrupt change in existence status of:
    1. Occupation relation from not existing, to existing, to not existing again.
2. Initiates abrupt transformations in pattern of extensional relations between the moving unit and the spatial place. (Organizational initiation—extrinsic consequence—existential-pathway-transformational-development).

First Case: Unit passing by a spatial place, and passing through the point of shortest distance relation, the transformation point.

1. Initiates abrupt change in manner of change of pattern of extensional relations, with components transforming from decreasing to increasing and from increasing to decreasing.

Second Case: Unit moving directly at the spatial place and passing through it, thus passing through the point of least (zero) distance relation, the transformation point.

1. Initiates abrupt change in manner of change of pattern of extensional relations, with components transforming from decreasing to increasing, from one orientation to the opposite orien-

tation, and from nonexistence to existence to nonexistence.

Initiation situation—Motion of a unit in relation to a static unit.

1. Initiates change in extensional relations between the moving unit and the static unit. (Organizational initiation—extrinsic consequence)

Change in:

1. Distance relations;
2. Direction relations, and;
3. Positional orientation relations.

2. Initiates change in pattern of extensional relations between the moving unit and the static unit. (Organizational initiation—extrinsic consequence—new pattern of material organization)

Pattern of Extensional Factors

First level pattern
Extensional pattern of coexistent spatial places.

(Immaterial extensional pattern differentiated locationally and by the unique self-identity of spatial places.)

1. Extensional pattern of adjacent spatial places.

    (Adjacent relations and direction relations).

2. Extensional pattern of nonadjacent spatial places.

    (Distance relations and direction relations).

Second level pattern
Extensional pattern of coexistent parts of a primal unit of matter.

> (Material extensional pattern differentiated locationally and by the unique self-identity of the parts of a primal unit.)
>
> 1. Extensional pattern of adjacent parts of a primal unit of matter.
>
>   (Adjacent relations, contact relations, and direction relations).
>
> 2. Extensional pattern of nonadjacent parts of a primal unit of matter.
>
>   (Distance relations and direction relations).

Third level pattern, transitional level
Extensional pattern of coexistent primary components, space and one static unit of matter.

> (Extensional pattern differentiated in part materially by a unit of matter and in part immaterially by location and by the unique self-identity of spatial place.)
>
> 1. Extensional pattern between a unit of matter and adjacent spatial places.
>
>   (Adjacent relations and direction relations).
>
> 2. Extensional pattern between a unit of matter and nonadjacent spatial places.
>
>   (Distance, direction, and positional orientation relations).

Fourth level pattern
Extensional pattern of coexistent primary components, two static units of matter.

(Material extensional pattern differentiated by extensionally limited units.)

1. Extensional pattern of adjacent primary components, two static units of matter;

    (Adjacent relations, contact relations, direction relations, and positional orientation relations).

2. Extensional pattern of nonadjacent primary components, two static units of matter.

    (Distance, direction, and positional orientation relations).

Simultaneity

First Form
Simultaneity of cases of continuing-existence—simultaneity of:

1. The continuing-existence of the coexistent parts of space;
    1. Simultaneous occurrence of the continuously new current part of the continuing-existence of a spatial place with the continuously new current part of the continuing-existence of other spatial place;
    2. Simultaneous occurrence of the noncoexistent-sequential-difference of the continuing-existence of a spatial place with the noncoexistent-sequential-difference of the continuing-existence of other spatial place.

2. The continuing-existence of the coexistent parts of a primal unit of matter;
    1. Simultaneous occurrence of new part of the continuing-existence of one part of a primal unit with new part of the continuing-existence of other parts of the unit;
    2. Simultaneous occurrence of the noncoexistent-sequential-difference of the continuing-existence of one part of a primal unit with the noncoexistent-sequential-difference of the continuing-existence of other parts of the unit.
3. The continuing-existence of matter with the continuing-existence of space;
    1. Simultaneous occurrence of new part of material continuing-existence with new part of spatial continuing-existence;
    2. Simultaneous occurrence of the noncoexistent-sequential-difference of material continuing-existence with the noncoexistent-sequential-difference of spatial continuing-existence.
4. The continuing-existence of motion with the continuing-existence of matter;
    1. Simultaneous occurrence of new part of motion continuing-existence with new part of material continuing-existence;
    2. Simultaneous occurrence of the noncoexistent-sequential-difference of motion continuing-existence with the noncoexistent-sequential-difference of material continuing-existence.

5. The continuing-existence of motion with the continuing-existence of space;
    1. Simultaneous occurrence of new part of motion continuing-existence with new part of spatial continuing-existence;
    2. Simultaneous occurrence of the non-coexistent-sequential-difference of motion continuing-existence with the noncoexistent-sequential-difference of spatial continuing-existence.
6. The continuing-existence of one unit of matter with the continuing-existence of another unit;
    1. Simultaneous occurrence of new part of the continuing-existence of one unit of matter with new part of the continuing-existence of another unit;
    2. Simultaneous occurrence of the non-coexistent-sequential-difference of the continuing-existence of one unit of matter with the noncoexistent-sequential-difference of the continuing-existence of another unit.

Second Form

Simultaneity of motion with cases of continuing-existence—simultaneity of:

1. The continuous aspect of ongoing motion with the continuous aspect of motion continuing-existence;
    1. Simultaneous occurrence of new part of ongoing motion with new part of motion continuing-existence;
    2. Simultaneous occurrence of the noncoexistent-sequential-difference of motion with the noncoexistent-sequen-

tial-difference of motion continuing-existence;

2. The continuous aspect of ongoing motion with the continuous aspect of material continuing-existence;
    1. Simultaneous occurrence of new part of ongoing motion with new part of material continuing-existence;
    2. Simultaneous occurrence of the noncoexistent-sequential-difference of motion with the noncoexistent-sequential-difference of material continuing-existence;
3. The continuous aspect of ongoing motion with the continuous aspect of spatial continuing-existence;
    1. Simultaneous occurrence of new part of ongoing motion with new part of spatial continuing-existence;
    2. Simultaneous occurrence of the noncocxistent-sequential-difference of motion with the noncoexistent-sequential-difference of spatial continuing-existence.

Third Form
Simultaneity of motion with changes of extensional relations between a moving unit and spatial place—simultaneity of:

1. The continuous aspect of ongoing motion with the continuous aspect of change of occupation/location, direction, distance, and positional orientation relations;

1. Simultaneous occurrence of new part of ongoing motion with new part of the change of extensional relations;
2. Simultaneous occurrence of the noncoexistent-sequential-difference of motion with the noncoexistent-sequential-difference of the change of extensional relations.

Fourth Form
Simultaneity of motion with changes of the pattern of extensional relations between a moving unit and spatial place—simultaneity of:

1. The continuous aspect of ongoing motion with the continuous aspect of changing pattern of extensional relations;
    1. Simultaneous occurrence of new part of ongoing motion with new pattern of extensional relations;
    2. Simultaneous occurrence of the noncoexistent-sequential-difference of motion with the noncoexistent-sequential-difference of changing pattern of extensional relations.

Unidirectionality

First Form
Unidirectionality of continuing-existence, of:

1. Space;
2. Matter, and;
3. Motion.

Second Form
Unidirectionality of motion.

Third Form
Unidirectionality of the change of extensional relations between a moving unit and spatial place, of:

1. Occupation/location relations;
2. Direction relations;
3. Distance relations, and;
4. Positional orientation relations.

Fourth Form
Unidirectionality of the change of pattern of extensional relations between a moving unit and spatial place.

Fifth Form
Unidirectionality of the transformation of extensional relations between a moving unit and a spatial place when the unit passes through a transformation point.

> First Case: Unit passing by a spatial place, and passing through the point of shortest distance relation, the transformation point.
> 1. Unidirectionality of abrupt changes in the manner of change of:
>   1. Distance relation, from decreasing to increasing;
>   2. Direction relation, from increasing rate of change to decreasing rate;
>   3. Positional orientation relation, also from increasing rate of change to decreasing rate.

Second Case: Unit moving directly at the spatial place and passing through it, thus passing through the point of least (zero) distance relation, the transformation point.

1. Unidirectionality of abrupt change in the manner of change of:
    1. Distance relation, from decreasing to nonexistence, then reoccurrence and increasing.
2. Unidirectionality of abrupt change in orientation of:
    1. Direction relation from existing, to not existing, to existing again but with opposite orientation;
    2. Positional orientation relation from that prior to passing through the transformation point to the opposite, subsequent to passing through.
3. Unidirectionality of abrupt change in existence status of:
    1. Occupation relation from not existing, to existing, to not existing again.

Sixth Form

Unidirectionality of the transformation of the pattern of extensional relations between a moving unit and a spatial place when the unit passes through a transformation point.

First Case: Unit passing by a spatial place, and passing through the point of shortest distance relation, the transformation point.

Unidirectionality of abrupt transformation of uniform pattern of noncoexistent-sequential-difference of changing pattern of extensional relations to a different uniform pattern of noncoexistent-sequential-difference of changing pattern of extensional relations, which continues without further transformation.

> (Involves unidirectionality of abrupt changes in manner of change).

Second Case: Unit moving directly at the spatial place and passing through it, thus passing through the point of least (zero) distance relation, the transformation point.

> Unidirectionality of abrupt transformation of uniform pattern of noncoexistent-sequential-difference of changing pattern of extensional relations to a different uniform pattern of noncoexistent-sequential-difference of changing pattern of extensional relations, which continues without further transformation.
>
>> (Involves unidirectionality of abrupt changes in manner of change, in orientation relations, and in existence status).

Seventh Form
Unidirectionality of the change of extensional relations between a moving unit and a stationary unit, of:

1. Direction relations;
2. Distance relations, and;
3. Positional orientation relations.

Eighth Form
Unidirectionality of the change of pattern of extensional relations between a moving unit and a stationary unit.

Uniformity

First Form
Uniformity of unchanging spatial extension.

Second Form
Uniformity of the change of continuing-existence, of:

1. Space;
2. Matter, and;
3. Motion.

Third Form
Uniformity of the change that is undisturbed motion.

Fourth Form
Uniformity of the change of extensional relations between an undisturbed moving unit and spatial place, of:

1. Occupation/location relations;
2. Direction relations;
3. Distance relations, and;
4. Positional orientation relations.

Fifth Form
Uniformity of the change of pattern of extensional relations between an undisturbed moving unit and spatial place.

Sixth Form
Uniformity of the change of extensional relations between an undisturbed moving unit and a stationary unit, of:

1. Direction relations;
2. Distance relations, and;
3. Positional orientation relations.

Seventh Form
Uniformity of the change of pattern of extensional relations between an undisturbed moving unit and a stationary unit.

Glossary

change—An occurrence of noncoexistent-sequential-difference. The difference that occurs with change is sequential, with the parts having no aspect of coexistence. It is not possible to simultaneously observe or measure two sequential parts of an ongoing change. When one part is there, when it exists, the prior part has ceased to exist and the following part has not yet come into existence. Developed change is the transition from one state of an individual factor or situation at one particular part of its continuing-existence to a different state at a different part of that continuing-existence. Change is also the transition of a factor or situation from nonexistence to existence or the transition from existence to nonexistence.

There are two known foundational sources of change, continuing-existence and motion. Because space is the primal existential context for all else that exists, the continuing-existence of space is the existential context for the continuing-existence of all else that exists. All forms of change occur with and conform to the change that is the continuing-existence of space. The change that is the continuing-existence of space is the change that is time.

Motion takes time to occur, and in that manner it is existentially-dependent on the continuing-existence of space. All other forms and cases of change, that we are able to analyze for their foundations, are existentially based on motion. The differences between these other forms of change are due to the particular mix of other factors that play roles in relation to spatial continuing-existence and motion. With initiation situation, the initiators are coexistent with other factors extrinsic to the initiators that play roles in the nature of the consequences. With organizational initiation, a form of initiation situation, the extrinsic factors play organizational roles in determining the nature of the consequences. (*See also* organizational initiation)

Emergence, cause, and through-flow are particularly important stages in the development of change. With emergence, the motion of matter changes the extensional relations between units of matter, thus changing the pattern of material organization. With cause in the form it has at its development-of-origin, the push emerging from the collision relation forces pattern of material organization change. With the through-flow situation, the flow of energy alters the organization of matter, and the organization of matter alters the flow of energy.

Change codevelops with newness in that there is an aspect of newness that occurs with any change. Change also codevelops with consequent-existence in that change is a universal consequent aspect of the occurrence of consequent-existence. Change codevelops with the initiators in that they initiate change by way of consequent-existence. All change is determinate and occurs by way of existential-pathway-development.

Change is noncoexistent-sequential-difference, wherein the parts of the sequence are not coexistent. It is distinct from coexistent-sequential-difference, such as the alternating sequence of stripes on the side of a zebra, where the stripes are coexistent. The factor, change, does not play a role in coexistent-sequential-difference. There is no change from one stripe to another, just sequential difference, coexistent-sequential-difference.

change development—Change development is change with enhancement. The enhancement can be simple or complex, as simple as the universal quantitative increase of time that has occurred, or as complex as the emergence of life.

Change is an occurrence of difference, existentially-dependent, sequentially noncoexistent difference. Change development is the transition with enhancement from one state of an individual factor or situation at one particular part of its continuing-existence to a different state at a different part of that continuing-existence. The states or stages of change development are noncoexistent. They are not, and cannot be, coexistent.

Change development is determinate. Existence initiates existence. Existence initiates change. It does so by way of determinate consequent-existence. Existence determines existence. Existence determines change. The existence and nature of that which goes before determines the existence and nature of that which follows by way of change development.

Change development originates with the change that is spatial continuing-existence. The first stage of change provides an existential context for all following stages. Change development again occurs with the continuing-existence of matter, and this change conforms to that of space. Change develops again with motion. The change that is motion interrelates extensional development and change development into a unity, into a single multifactor, multidevelopmental situation. Motion developmentally interrelates together the unchanging coexistent-sequential-difference of extensional development with the changing noncoexistent-sequential-difference of change development. Change next develops with the relative motion between units of matter. This is the development-of-origin for emergence. Apparently, the rest of change development occurs as the development of emergence.

Change development is characterized by existential-pathway-development. Existence initiates and determines existence and change. That which goes before determines that which follows. Change development is the transition from one state of a factor to a different, following, noncoexistent state of that factor, the factor maintaining at least some of its individual self-identity during the transition. The change that occurs follows the continuing-existence of the factor. There is a connection of existence from the prior state to the following state. The sequence of change, of development, follows this pathway of continuing-existence—existential-pathway-development.

Existential-dependency is also characteristic of change development. Initiators initiate change. They do so by way of determinate consequent-existence. There is an existential-pathway-development from initiator to consequent. The existence of the consequent is dependent on the existence of the initiator. With change development, following stages are existentially-dependent on prior stages of the existential-pathway-development.

All of reality is continuously going through change development, noncoexistent sequential enhancement, in the manner of continuing-existence, the continuous addition of new part of continuing-existence. Virtually all of reality is continuously going through change development in the manner of motion, the continuous addition of new part of ongoing motion. And virtually all reality is in a process of continuous emergence, the change development of the initiator motion in relation to pattern of material organization, the continuous occurrence of newly emergent pattern.

Situations can develop through the enhancement of additional quantity, such as the addition of units to a group, or through the enhancement of increased complexity through reorganization of a group. Situations can also lose units, or become less complex through reorganization. A situation can lose these modes of enhancement. Nonetheless, the situations still develop in that they always have an aspect of development based on the simpler universal modes of enhancement, based on continuing-existence, motion, and emergence.

coexistence—Coexistence is a situation in which two or more factors exist in space, anywhere in space, and do so during the same part of the continuing-existence of space. Everything that exists in infinite space, in all reality, is coexistent. That which exists in space is coexistent with space. The continuing-existence of that which exists in space occurs simultaneously with the continuing-existence of space. The continuing-existence of a factor that exists in space during a particular part of spatial continuing-existence occurs simultaneously with the continuing-existence of a different factor that also exists somewhere in space during the same part of spatial continuing-existence.

The term coexistent is not restricted to situations where the primary components are primal-forms-of-existence or material objects. For example the term can refer to the coexistence of the three-dimensionality of space and the three-dimensionality of matter—coexistent factors of dimensionality.

Any factor that exists and that can play a role as a primary component of a coexistence has some aspect of continuing-existence involved with its mode-of-being. Such a factor always exists at constantly new part of its continuing-existence. Coexistence situations derive an aspect of continuing-existence from that of their primary components, and thereby an aspect of continuous newness. A coexistence situation between primary components continues to exist, and exists always at constantly new part of that continuing-existence.

The ever-new aspect of the continuing-existence of a primary component provides the existential context for change of, or in relation to, that primary component. The derived, existentially-dependent, ever-new aspect of coexistence relations provides the existential context for change in intrinsic aspects of coexistence situations. It also provides the existential context for changes that are associated with coexistence situations. For example, a moving object occupies a sequence of coexistent spatial place. At a prior part of the ongoing continuing-existence of the coexistence of two sequentially organized spatial places, the object first occupies one of the pair of spatial places but not the other, and then, at a following part of the continuing-existence of the coexistence of the two places, the object occupies the other place and is no longer at the first place. Change occurs associated with each of the two primary component places, it does so at different parts of the continuing-existences of the places and thus at different parts of the continuing-existence of the coexistence.

coexistent-sequential-difference—Sequential organization in which the components of the sequence are coexistent. Coexistent-sequential-difference of any form or stage of development is existentially-dependent on the coexistent-sequential-difference of spatial extension. Spatial places have unique self-identity based on their existence and their individual unique locations in infinite space. In a sequence of spatial places, each is existentially unique and thus sequentially different. Developed forms of coexistent-sequential-difference are based on matter, on the individual self-identity of the coexistent units in the sequence and on their organization in space, their coexistent occupation of a sequence of spatial places.

Change, which occurs as noncoexistent-sequential-difference, plays no role in the intrinsic nature of coexistent-sequential-difference.

combinatorial enhancement—Combinatorial enhancement is the occurrence of relations between components of a coexistence situation. If there is coexistence, there is relation. The relation is in addition to and existentially-dependent on the existence of the primary components of the coexistence.

The occurrence of combinatorial enhancement is why the whole is greater than the sum of the parts. A simple numerical summation of the components of a group does not include the relations between those components. In any group of three or more, the sum of the relations is greater than the sum of the components. Even with just spatial relations, distance and direction relations, the sum of the relations increases faster with additional components than does the sum of the components. Combinatorial enhancement has a major role in the origin of complexity.

Since combinatorial enhancement is relation that occurs with coexistence, combinatorial enhancement does not require newness, just coexistence. There are two foundational cases of combinatorial enhancement, one with space and one with matter, with the material case existentially-dependent on the spatial case for a place-to-be. There is the combinatorial enhancement that occurs with the coexistence relation of spatial places, the direction, distance, and sequential-difference relations between those places. And there is the combinatorial enhancement that occurs with the coexistence relation of parts of a primal unit, the contact, direction, distance, and sequential-difference relations between those parts.

Combinatorial enhancement is a factor of change development when a relation is new as a result of any form of new combining. Combining is any of various modes of togetherness that have factors beyond coexistence that result in combinatorial enhancement. For there to be combining beyond mere coexistence there must be a factor of combination, which foundationally are relations of occupation, location, interaction, new coexistence, and coherence. When factors newly combine, newly occurring relations occur between them. These relations are new factors, and because they are new, they are enhancements of the situation, enhancements due to new combining—combinatorial enhancements.

When combinatorial enhancement results from new combining, it does so as a result of a role of sequential enhancement. New combining is change development, and change development always has a role for sequential enhancement. Combinatorial enhancement can be a consequence of sequential enhancement. Sequential enhancement, however, does not always result in a case of combinatorial enhancement. It does so only when there is a factor of combination playing a role in the situation.

At the foundational stage of combinatorial enhancement, the coexistence of distinct spatial places, the components of the coexistence are already there together, eternally so. Newness, sequential enhancement, and change development play no roles. At later stages change development and its factors result in combinatorial enhancement. But in many of these situations, the components of the coexistence are already coexistent, as in the cases of new occupation/location relations of a moving unit of matter to space, and the physical contact interaction of colliding units. In these situations there is a factor of combination that plays a role beyond that of coexistence, such that the components are more together, more combined, than they are by way of simple coexistence.

Different kinds of factors can combine and have consequent relational enhancements. The manner of combining and the various differences in the intrinsic nature of the primary components determine the nature of the relations between the coexistent primary components. Therefore the nature of the enhancements that occur with coexistence varies from situation to situation, and these enhancements can be recognized as secondary coexistence factors.

Combinatorial enhancement occurs developmentally prior to emergence, as the coexistence between spatial places. After a few stages of development, it becomes a major factor in one of the two basic pathways of universal emergence—emergence from combining. Combinatorial enhancement at this stage is the consequent of the process of emergence wherein factors of combining play roles.

compositional subunits—The building blocks of coherent structure, for example elementary particles, atoms, molecules, bricks, logs, units of lumber, tires, window panes, bones, cells, organs, and on and on. Compositional subunits are not broken chunks of substance or material such as pieces of broken glass, (unless, of course, they are reused as in a mosaic or in asphalt roadway material).

compound change—Change based on both continuing-existence and motion. Compound change is founded on motion and develops with the initiators as more and more factors play roles. All stages of the development of change beyond simple change are based on compound change.

consequent-existence—The existence of a factor as a consequence of the prior or simultaneous existence of another factor. The development-of-origin of consequent-existence is the initiation of the continuing-existence of space as a consequence of the existence of space. Major foundational stages of its development occur with the existence of matter, with the existence of motion, and with the developments-of-origin of emergence and cause. With consequent-existence, existence initiates existence—with existence initiating continuing-existence, with the existence of motion initiating ongoing motion, with components and their interrelations initiating emergent pattern, and with matter in motion in material contact interaction initiating cause and effect. Consequent-existence develops by way of differences in the way factors initiate the existence of other factors.

Consequent-existence is determinate. That which goes before determines, by way of the ongoing continuance of what it is, the existence and nature of what follows.

continuance-of-being—Continuing-existence.

continuing-existence—For there to be existence, there must be some continuance of existence. To not have any continuance of existence is to not be there at all.

When something exists, and when there are no other factors playing roles that interfere with that existence, that something will simply continue to exist. It exists, and it continues to exist. Existence initiates continuing-existence. As something continues to exist, new part of that continuing-existence continuously comes into existence. As new part of continuing-existence comes into being, the prior part ceases to exist. The parts of continuing-existence are noncoexistent. When a particular part of continuing-existence is occurring, the following part is not there, it has not yet come into existence. When that following part is occurring, the prior part is no longer there.

Continuing-existence is a form of noncoexistent-sequential-difference, a form of continuous change with only the current part in existence. Existing in the form of continuous change, the current part has no dimension. Dimensionality plays no role in the nature of continuing-existence.

The continuous initiation of new part of continuing-existence results in a sequence of noncoexistent-sequential-difference with after and before relations between the noncoexistent parts. Because continuing-existence is nothing more that continuance-of-being, its occurrence is uniform and unidirectional.

Spatial continuing-existence is the foundational case of continuing-existence to which all other cases conform.

determinate-reality—That reality is determinate, and cannot be otherwise, is due to several factors and their roles in the foundational development of reality. These factors range from existence itself to through-flow. They interrelationally develop with the fundamental development of reality, each factor continuing to play its role with the accumulation of additional factors until through-flow exists as all these factors playing their roles in concert.

Reality is that which exists, and existence sets the foundation of determinate-reality by setting self-identity. What exists is what exists—what exists is itself, with all its intrinsic qualities. What exists is what is-there to play further roles in the nature and development of reality, doing so by way of its self-identity, by way of what it is.

That which exists continues to exist. Existence initiates continuing-existence. The continuance is a consequence of the existence. This is consequent-existence, the existence of one factor as a consequence of the existence of another factor. With continuing-existence, it is a matter of the continuance of what is there. What is there simply continues to be there. Thus, what is there before determines by way of its continuing-existence what it is that continues to be there. This includes both the existence and intrinsic nature of what is there. Thus, the existence and intrinsic nature of what is there before determines, by way of the continuing-existence of its self-identity, the existence and intrinsic nature of what continues to be there. Consequent-existence is determinant.

Motion exists, and thereby it is what it is. Motion is matter passing through space. It is a form of change, with the moving matter passing through and momentarily occupying a sequence of spatial place. As the matter occupies a place, it momentarily shares the extensional relations that place has with all the rest of spatial place. As the moving matter occupies the sequence of spatial place, there occurs a sequence of changing extensional relations between the moving matter and space.

The motion exists, and continues to exist. It continues to be itself. It continues as motion, with its intrinsic qualities such as its speed and direction. The continuance of the motion, based as it is on continuing-existence, is determinate. The continuing-existence of the motion is a consequence of the existence of the motion, a case of consequent-existence. With continuing-existence there is a continuance of what exists, along with its intrinsic qualities. The change from what exists before to what exists following is no more than a change from one part of continuing-existence to a following part, an intrinsic change.

Motion, however, occurs in relation to space. There are changes of the occupation and extensional relations of the moving matter. These changes are consequences of the motion. Which places are occupied and which extensional relations are thereby shared are consequences of the direction of the motion, and the rate at which the changes take place is a consequence of the speed. The motion and its qualities determine what changes occur. This is consequent-existence, wherein what goes before determines what follows. The motion, though, takes place in relation to an extrinsic context, space. It is the interrelation of the existence and intrinsic qualities of the motion with the existence and intrinsic qualities of that context that fully determine the existence and nature of the consequent changes.

When units of matter exist in space together, there is a pattern of extensional relations among them, a pattern of material organization. When the units move, the extensional relations change, and new pattern of material organization comes into existence—it emerges. The factors of determinate-reality of prior stages are here playing their roles, but now they are doing so in the context of a group of material units. The changes of extensional relations between the units, and the change of material organization, are consequences of the motion of the units in relation to the group. Again it is determinant consequent-existence in that the existence and intrinsic nature of all the factors playing roles in the situation determine, by way of the continuing-existence of what they are, the existence and intrinsic nature of the following situation. Emergence, here at its development-of-origin, is determinate.

The next development of the determinate aspect of reality takes place in collision situations. Collisions develop from a basic emergence situation, the simplest case being one unit moving on a path directly at another, stationary, unit. As just described, all aspects of this situation are determinate, with the changing relations of the units determined by the motion of the one unit in relation to the location of the stationary unit. A collision, then, is a determinate consequence of the continuing-existences of the factors involved. Upon the occurrence of adjacent relation and contact between the units, the continuing-existence of the motion, the continuance of its self-identity, its intrinsic nature to continue on as motion, has the determinate consequence that the moving unit presses against the blocking stationary unit. The emergence of this pressure, this push, is the development-of-origin of cause.

While it is the continuance of the direction of the motion in relation to the continuance of the stationary unit at its blocking location that leads up to the adjacent relation, the contact, and the push, it is the existential quantity of the moving matter and its speed in relation to the existential quantity of the blocking unit that determines how much pressure there is. It is determinate consequent-existence, the existence and nature of what goes before determining, by way of the continuing-existence of the involved interrelating factors, the existence and nature of what follows.

The blocking unit prevents to various degrees the onward motion of the other unit. That unit loses motion as it pushes on the stationary unit, and when the stationary unit receives the push it begins to move. It acquires motion. In the collision situation, the moving unit causes the stationary unit to move. There is a transfer from one unit to the other in what is the development-of-origin of through-flow. As in the prior stages of the collision, the nature of the factors going into the through-flow situation determine, by way of what they are, the nature of the consequent factors.

Reality is determinate because of what it is foundationally. Reality is determinate because (1) existence sets self-identity, (2) existence initiates continuing-existence, (3) with continuing-existence what is there determines what continues to be there, (4) motion is a form of change, and with the continuing-existence of motion, motion initiates further motion, (5) motion initiates changes in occupation and extensional relations between a moving unit of matter and spatial place, (6) motion initiates (a) changes in extensional relations between units of matter, (b) change in the material pattern of organization, and (c) the origin of emergence, (7) all these factors together in a collision situation initiate the origin of cause, and (8) cause initiates effect, originating through-flow wherein an energetic transfer from unit to unit plays a role in the nature of the consequences. Reality is determinate because in all these stages there is determinate consequent-existence, and because in all stages where change plays a role (2-8), the existence and nature of what goes before determines, by way of the continuing-existence of self-identity, the existence and nature of what follows. The determinate aspect of reality continues to develop as additional factors play roles, and can become seriously complex as with the development of an organism from egg to adult.

Determinate-reality should not be confused with epistemological determination. The latter is the mental process of figuring something out, for instance by thinking about it, discussing it, or using logic or math. While determinate-reality is a universal feature of reality, epistemological determination occurs only where epistemological factors play roles, and in the universal context, that is an exceedingly rare occurrence—only where sentient beings think.

development—A development is a difference from one place, part, state, condition, or situation to another involving some form of enhancement. The enhancement can be starkly simple or exceedingly complex, and develops overall from the one to the other. There are two primary forms of development, extensional development and change development. The foundation of extensional development is the extension of spatial place, and the development consists of differences in various factors from spatial situations of less extension to spatial situations of greater extension. The foundation of change development is spatial continuance-of-being, the primal origin of change, and the development consists of the sequence of continuous change from one part of continuing-existence to the next part.

With extensional development, at its foundational level as an aspect of the nature of spatial place, the parts of the continuous sequential-difference are coexistent and unchanging—coexistent-sequential-difference. With change development, the parts of the continuous sequential-difference are noncoexistent and change plays a role—noncoexistent-sequential-difference.

The extensional development of spatial place and the change development of spatial continuing-existence provide the existential contexts for all other forms of extensional development or change development. All developed forms of development conform to the foundational forms provided by space.

There is an aspect of continuity with these two forms of development. Each prior stage develops directly into the following stage. The existence of that which is prior develops directly into the existence of that which follows. This existentially continuous, sequentially connected development is existential-pathway-development. It occurs intrinsically with individual factors, and also with situations as the combined interrelational existential-pathway-developments of the factors and components of a situation.

Factors often occur in simple form in simple situations where few factors are playing roles, and occur in more complex form in more complex situations where larger numbers of factors are playing roles. This is factor development. There are two basic ways in which it occurs—(1) by way of existential-pathway-development and (2) by way of nonpathway factor development. With existential-pathway-development, factor development can occur in two ways, (1) a stage of the development of a factor develops directly into another stage, a direct transformation of one stage into a different stage, and (2) there are intervening stages of the existential-pathway-development between the first stage of the factor and the occurrence of the other stage.

The second basic way in which factor development occurs is by way of nonpathway factor development. Nonpathway factor development is the development of a factor from one change developmental pathway to another, which is in some manner unrelated to the first. A factor can occur in simpler form in one developmental pathway, and occur also in a more complex form in a different more complex developmental pathway. Continuance and sequential-difference play roles in the nature of both extensional development and change development, but not in nonpathway factor development.

Development is difference with enhancement. There are two foundational forms of enhancement, sequential enhancement and combinatorial enhancement. They both play roles in extensional development and in change development, but neither plays a role in nonpathway development. The basic form of sequential enhancement is increasing quantity, while the basic form of combinatorial enhancement is the occurrence of relation in coexistence situations. These two forms of enhancement developmentally interrelate.

The two basic forms of development, extensional and change, also developmentally interrelate. The foundational form of change, continuing-existence, does not involve extensional-development, but motion does. Motion is matter passing through space, continuously and sequentially changing the spatial place the moving matter occupies. Motion occurs in relation both to the extension of space and to the change that is continuing-existence. Because all developed forms of change involve motion in one way or another, developed forms of existential-pathway-development involve the interrelation of extensional and change developments.

Change development is due to the roles of the initiators. These are factors that make change happen. They do so by way of consequent-existence. With consequent-existence, the existence of one factor has the consequence that another factor comes into existence. Because the initiators exist, various forms of change exist, each with associated developments. The existence of space and matter initiate the change that is continuing-existence. Motion initiates ongoing motion. The motion of matter in relation to other matter initiates the change that is emergence. Moving matter in collision relation with blocking matter initiates caused change. And in the through-flow situation, the flow of energy alters the organization of matter, and the organization of matter alters the flow of energy, resulting in the major portion of the change that humans experience. Each of these stages of the development of initiation, consequent-existence, and change development is composed of the prior stages plus the factors that make each stage a development form the prior stage.

What have been described here are the foundational aspects of development. Development develops into complex initiation situations such as biological evolution, ontogeny, and thinking.

developmental transformation point—*See* transformation point

development-of-origin—(1) Any development at which a factor comes into existence. Thereafter the factor can play roles in existential-pathway-development and in situation development. (2) In the general development of reality, the stage at which a factor first occurs. Usually a factor first occurs in simple form, and becomes more complex with factor development. (3) A difference or change in an existential-pathway-development as a consequence of which a factor first occurs in that particular developmental pathway. (4) In a single existential-pathway-development, a particular type of factor can originate at two different stages of the pathway development. Each case of the factor in that pathway has its individual development-of-origin. For example, the successive generation of branches along the trunk of a conifer such as a Norfolk Pine. Each set of branches has its individual development-of-origin, as does each branch.

Because in many kinds of developmental pathways, conditions vary at different stages of pathway development, a particular type of factor can originate at almost any stage of its own factor development, depending on the nature of the pathway. A particular type of factor can also originate at different stages of pathway development in different kinds of developmental pathways.

Here are two examples of developments-of-origin. (1) The situation of a unit moving in space is the first stage of organizational initiation. The development from a unit stationary in space to a unit moving in space is the development-of-origin for organizational initiation. (2) The situation of a unit moving in relation to a stationary unit is the first stage of emergence. The development from the situation of two nonadjacent units stationary in space to the situation in which one of the units is moving is the development-of-origin for emergence.

direct development—(see also indirect development) With direct development there is a direct existential-dependency relation between the consequent and the initial condition, without intervening stages or levels. With direct factor development there is a direct existential-dependency relation between a prior stage of a factor and a following stage. For example, a particular case or instance of a factor undergoes a change to a subsequent developmental stage with the physical and organizational basis of the prior stage still constituting the major physical and organizational basis of the enhanced following stage.

Existential-pathway-development occurs by way of direct development.

distance—Amount or quantity of linear space, spatial place, extension, or coexistent-sequential-difference between two points, places, or locations.

emergence—Foundationally, universally, and at all stages of its development, emergence is the creation of newly existing patterns of material organization as a consequence of the motion of units of matter relative to one another. This is the core nature of emergence, the feature that makes it emergence—in every case, no matter how simple the situation, no matter how complex. It can occur in situations with as few as two units, or as the entire infinite universe.

At its development-of-origin, emergence occurs as a noncausal simple form of determinate consequent-existence, becoming causal at a later stage when material contact interaction first plays a role in the process. At the stages where there is no contact between units, the structural logic that dominates the nature of emergence is that of space, time, motion, and the relations between them. With stages in which material contact interaction does play a role, the intrinsic qualities of the units in large part determine the structural logic of the creative process and the nature of the resulting pattern of material organization.

existential context—This relation occurs when one factor provides a situation that allows for the existence of another factor that could not have existence without the provided existential context. The foundational examples are the place-to-be in which matter exists that is provided by immaterial spatial place, and the universal context of change provided by spatial continuing-existence for the change of material continuing-existence and all other forms of change.

existential dependency—This is a relation in which the existence and nature of one factor is dependent on the existence and nature of another factor. There are three basic forms, (a) existential-context-dependency, (b) the existential-pathway-developmental-dependency that occurs with change development, and (c) hierarchic-existential-dependency. The foundation of existential-context-dependency is the manner in which the existence of everything other than space is dependent on the existential context provided by space. The foundation of existential-pathway-developmental-dependency occurs with the continuing-existence of space, and develops with change development. This form of dependency is the manner in which the existence and nature of any stage of a change existential-pathway-development is dependent on the prior development of the situation, on the history of the situation. In its early stages, existential-pathway-developmental-dependency codevelops with initiation. Hierarchic-existential-dependency is the manner in which the whole is dependent on the parts. The existence and nature of the whole is dependent on the existence, nature, and manner of togetherness of the parts. Other than the role of the way in which they came together, the manner of the togetherness of the parts is dependent on the nature of the parts. One form of hierarchic-existential-dependency is structural existential-dependency, wherein the nature of the whole is dependent on the structural relations of the parts, as is so evident with the Eiffel Tower.

existential extension—The extension of space or of matter that must exist for there to be space or matter.

existential factor—A factor whose role is required for there to be existence. Existential factors are factors that constitute the existence of something. A primal existential factor plays a role in the foundations of reality, such as the extension, voluminality, and continuing-existence of space or matter. More generally, a factor required for the existence of the intrinsic nature of something, as oxygen for the existence of water, H_2O. (Consequent-existence, self-organization, emergence, and cause, factors of the origins of things, are developmental factors rather than existential factors.)

existentially-dependent development—(see also existentially independent development) This occurs with the existential-pathway-development of change development. The existence and the role of a prior stage is required for the existence of a more developed stage. In general, change development occurs by way of determinate consequent-existence. The existence of the consequent situation is existentially-dependent on the existence of the initial situation.

More specifically, this is the dependency of developed stages of a factor on previous stages of that factor. A previous stage can be required as an earlier component in a creative process, and its current presence is no longer required. (Indirect) Its current presence can be required as a component of the developed stage. It can be required as a separate case of the occurrence of the factor as in a subcomponent of the developed case. (Indirect) Or the developed case can be the still existing prior case in changed, developed, form. (Direct)

There is a general existential-dependency relation between all that exists other than space and the existential context provided by space. This can occur more specifically with some factor developments. For example, the extension of matter is dependent on the place-to-be provided by the extension of space. (Indirect)

existentially independent development—(*see also* existentially-dependent development; nonpathway factor development) The existence and the role of a simpler or prior stage is not required for the existence of a stage of the development of a factor which is more complex than a separate, simpler, occurrence of the factor. Existentially independent development is disconnected development wherein there is no aspect of existential-dependency of a more developed form of a factor on a simpler form. Existentially independent development occurs because a factor or a pattern can often originate independently in different kinds of situations, in different kinds of developmental pathways.

existential-pathway-development—Occurs when the existence of that which is prior develops directly into the existence of that which follows. It is the existentially continuous, sequentially connected development of a particular individual factor or situation. It occurs with both extensional development and change development. With extensional development, it involves continuously more existential quantity. With continuously more spatial or material extension, there are accompanying developments between the parts of extension. Extensional existential-pathway-development can occur with material structure such as the additional structural relations that are there with increased extension through an already formed crystal.

As a characteristic of change development, existential-pathway-development is the existentially continuous, sequentially connected development of a particular individual factor or situation during its continuing-existence. The factor or situation can be as simple as space, or an elementary particle, or as complex as the human body, or the infinite universe. The existential-pathway-development that goes with change development is the progressive transformation of a situation—the continuing-existence of space, the growth of a crystal, the growth of a tree, the development of a forest ecosystem on the terrain left behind by a retreating glacier, the origin of a star and its family of planets and other orbitals.

All existential-pathway-developments with a role for change are cases of determinate consequent-existence. The existence and nature of what goes before in change development existential-pathway-development determines the existence and nature of what follows. This form of development is existentially-dependent in that the existence of a developed stage is dependent in one manner or another on the existence of a previous stage.

existential-pathway-transformational-development—Existential-pathway-development based on the role of a transformation point.

existential quantity—For there to be existence, there must be some amount or quantity of that which exists. If there is no quantity at all of something, that something does not exist. It will not, it cannot, be-there.

The existential quantity of space is, and must be, voluminal. This voluminality of space can be viewed, for practical purposes, as three-dimensional. In a mental model of space, picture a box-like area that has height, width, and length. Now it is not actually possible to compress space itself, but imagine with the mental model the top and bottom of the selected area coming together. Have them come together all the way, completely, so that there is no height left there. If there is no dimensionality left in the height, no extensional spatial place in the height dimension, there is not anything left there, no spatial place, to have dimensionality in width and length. If there is no existential quantity in one dimension, there is none there in the other dimensions. It is not possible for there to independently exist something that has only one or two dimensions.

The situation is the same for matter. To exist, matter must have some existential quantity, and that existential quantity must be voluminal. This is a requirement for the existence of substantiality. That which moves, can have a contact relation, that can block motion, that can push, that can form changeable pattern, and that can form coherent structure requires, as a factor of its existence, voluminal existential quantity.

Motion is existentially-dependent on matter. It is matter that moves. Any case of motion has an aspect of voluminality based on the voluminality of the matter that is moving. If there is no three-dimensional piece of matter there to move, there cannot be any motion there.

Existential quantity is also required for there to be the two foundational forms of noncoexistent-sequential-difference, that of continuing-existence and that of motion. For something to exist there must be some continuance of that existence, some existential quantity of continuing-existence. The same goes for motion itself. There must be some continuance of motion for there to be motion, some existential quantity of motion.

Just as there is a voluminal aspect of motion, the noncoexistent-sequential-difference of continuing-existence is always that of some voluminal form of existence.

For there to be existence, the amount of existential quantity required can be quite small. But the term does not refer only to the minimum necessary for existence. It also refers more generally to all the existential quantity of something that exists. Thus space, which is infinite, has infinite existential quantity, while a limited unit of matter has a specific limited amount.

extensional development—Extensional development is extension with enhancement. Foundationally the enhancement is simply more extension or more sequential-difference. There are two initial developmental pathways for extensional development. One is a codevelopment of extension with the three stages of the development of dimension—the one-dimensional extension of a linear location, the two-dimensional extension of a plane location, and the three-dimensional extension of three-dimensional spatial place. With this codevelopment pathway, the mode-of-being of each stage of dimensionality requires a larger quantity of extension than the previous stage—a whole new dimension of it. This is a limited development in that it can have only the three stages. A fourth stage is impossible because the three-dimensionality of spatial place fulfils all the possibilities of dimensionality, leaving no place for a fourth dimension to exist, and thus no place for a fourth stage of the codevelopment of extension and dimensionality.

The second basic developmental pathway for extensional development is that with any increase in the quantity of extension, at any stage of the development of dimensionality, there is a concomitant increase in quantity of sequential difference. With the increase in quantity of extension pathway, if a linear, plane, or three-dimensional location has more extension than another location, then that larger location has more sequential-difference.

Because space is an immaterial continuum, this developmental pathway is uniform, the greater the extension, the greater the sequential-difference.

Beyond increase in quantity of extension in relation to the development of dimension, and beyond increase of sequential-difference with increased extension, further stages of extensional development involve differences in distance and direction relations between spatial places that occur at various distances and directions from a third spatial place. More developments along this path involve differences in relations of area to circumference and volume to surface area that occur with increasing quantity of spatial extension.

Because space provides the existential context for all else that exists, and because to not exist in space is to not exist at all, that which exists other than space conforms to the factors of spatial existence. The extensional factors of that which exists conform to the extensional factors of space. Factors of motion and factors of material organization that involve factors of differences in extension conform to the extensional development of space. In this manner extensional development continues up through the material stages of the organization of reality, wherever extension plays a role, establishing there what can and what cannot exist.

factor—A factor is anything that exists and plays a role, whatever its mode-of-being—immaterial, material, organizational in space or in time, causal, or various simple or complex combinations. A factor of a situation is anything whose existence in the situation influences the nature of the situation. A combination of interrelated factors, because it exists, and because it can play a role that influences the nature of a larger situation, is a developed form of factor, a situation.

factor development—Factor development is the enhancement of a factor. Some factors are primally foundational, while other factors have developmental stages of origin. A change in an existential-pathway-development at which a factor originates is that factor's development-of-origin. Primally foundational factors do not have developments-of-origin. Developed forms of a factor are cases that are in one way or another enhanced forms of the factor beyond the form it has at its foundation or development-of-origin, whether they occur as a result of existential-pathway-development or nonpathway development.

indirect development—(*see also* direct development) With indirect development there is an aspect of existential-dependency between a prior stage of a factor and the following stage, but there are intervening developmental pathway stages or levels between them, or there are no intervening stages or levels but the mode-of-being of the prior stage of the factor does not play a role in the mode-of-being of the following stage.

initiation—Consequent-existence of change. Initiation is the source of change in change development, in existential-pathway-development. Nothing comes into existence, by any route whatsoever, without the role of initiation. Change by way of self-organization, emergence, and cause is derived from the roles of the initiators, from initiation.

With consequent-existence that which goes before determines that which follows. With consequent-existence, existence determines existence. With initiation, existence initiates existence, the existence of change, the existence of new part of noncoexistent-sequential-difference. Initiation is a consequence of the existence of space, of the existence of the primal form of matter, and of the existence of motion, and originates in these cases in determinate noncausal form. It does not become causal until the development-of-origin of that factor.

The development-of-origin of initiation in its most fundamental form is the initiation of the continuing-existence of space as a consequence of the existence of space.

Space exists, it is-there, and it continues to be-there, it continues to exist. The continuance is a direct consequence of the existence. It is simply existence continuing to be what it is—existence initiating existence by way of continuing to be existence. It is evident through observation that the parts of continuing-existence are not coexistent. Of spatial continuing-existence, only the present part exists, past part having ceased to be-there, and future part not yet having come into existence. The present part of this noncoexistent-sequential-difference of the continuing-existence of space is constantly changing, with new part continuously coming into being. That is, there is a continuous initiation of new part of continuing-existence. There is a continuous initiation of change.

Initiation occurs in a similar manner with matter. Matter exists, it is-there, and it continues to be-there, it continues to exist. Again, the continuance is a direct consequence of the existence, with continuous initiation of new noncoexistent part of the continuance as matter continues to be-there. There is here a continuous initiation of change from part to part of material continuing-existence.

The extension of spatial place provides an existential context for the existence of the extensional aspect of matter. Without the existential context of spatial place, matter could not exist. Spatial continuing-existence also provides an existential context, in this case for material continuing-existence. Without spatial continuing-existence, material continuing-existence could not occur. Thus, as matter continues to exist, it does so in the presence of, and in existential-dependency relation with spatial continuing-existence. In every case of material continuing-existence there are always two cases of continuing-existence occurring together. As the factor initiation develops from the spatial case to the material case, the spatial case is still there playing its role.

Space and matter are entirely different in the bases of their modes-of-being, space with its immateriality and matter with its substantiality. Yet each exists and initiates a case of continuing-existence. Motion, while being existentially-dependent on the existence of both space and matter, is still entirely distinct from either of them in its mode-of-being. But it too exists, and continues to do so. Motion, because it exists, initiates its own case of continuing-existence. Because of its existential-dependency relations, when there is initiation of motion continuing-existence, both developmentally prior cases are there playing their roles—three cases of continuing-existence developmentally related.

In addition to initiating a case of continuing-existence, motion initiates the continuance of motion. Because motion is itself a form of change, when it continues to exist there are two forms of change occurring. As a form of noncoexistent-sequential-difference, with only the current part in existence, there is with motion a continuous occurrence of new part, an initiation of new motion by way of motion continuing to exist, by continuing to be what it is.

Why motion exists at all is not known, and there may well be an additional form, or forms, of initiator and consequent initiation associated with its origin. Currently for the modern generalist mode, motion is just there. A generalist observes how it fits in with the rest of the factors of existence and organization, and works on from there.

The noncoexistent-sequential-difference aspect of motion cannot exist without the motion continuing to exist. The noncoexistent-sequential-difference of the motion itself occurs coexistent with and existentially-dependent on the noncoexistent-sequential-difference of motion continuing-existence. There are with motion the roles of four cases of noncoexistent-sequential-difference, three of continuing-existence and that of motion itself. With the development of the first four stages of initiation there is an accumulation of concurrent roles and existential-dependency relations.

The first four stages in the development of initiation are existential-pathway-developments from individual factors. In all four cases the consequences are intrinsic to the factors themselves—their continuing-existence, the continuance of the intrinsic nature of what they individually are. These four cases constitute the first of two general sequences of the development of initiation. The first sequence is characterized by existential-pathway-developments from individual factors, the three initiators, and has intrinsic consequences. These are cases of consequent-existence by way of existence only. The second sequence consists of all developments of initiation thereafter, the development of initiation situations. These are cases of consequent-existence by way of relations between the initiators and other factors. When initiators occur in relation to other factors, the consequences are extrinsic to the initiators themselves, generally consisting of changes in the relations between the initiators and those other factors, or of changes in relations between the various other factors involved in the situations.

The developmentally first stage of initiation situation, as currently understood, is the motion of a single unit of matter through space. During the continuing-existence of space, matter, and motion there are changes, initiated by the motion of the matter, in the occupation, distance, and direction relations between the moving unit and spatial place.

Continuing-existence and motion are intrinsically unidirectional—initiation is unidirectional. Because initiation of change is a form of consequent-existence, and because consequent-existence is determinate, initiation is determinate.

initiation situation—Consequent-existence of change by way of interrelations between the initiators and other factors.

initiator—A factor whose mere existence has the consequence that there is change, that there is noncoexistent-sequential-difference. There are three known initiators, space, the primal form of matter, and motion. They initiate the two known foundational forms of noncoexistent-sequential-difference, continuing-existence and motion. All three initiators initiate continuing-existence, while motion additionally initiates ongoing motion.

When initiators occur in relation to other factors, forming initiation situations, developed forms of change result, for example emergent change and caused change. The mere existence together of the factors that constitute the process of emergence has the consequence that creative change occurs, and the mere existence together of the factors that constitute cause has the consequence that caused change occurs. Sometimes initiation situations are so specific in form and so significant in the roles they play—emergence, the process, creates new pattern of material organization and cause creates effect—that they can be viewed as developed forms of initiators, and can be called initiators as long as their true status as initiation situations is kept in mind.

level of development—Hierarchic stage of development. Hierarchic organization is significant because, as a form of combinatorial enhancement, it greatly increases the number of patterns of organization that can exist. The level of development of hierarchic organization plays a role in how many different types of patterns of organization can exist in a particular situation. Reality is complexly hierarchic in nature, with diverse levels interrelating in seeming limitless patterns.

matter—That which exists that is not immaterial—that has substantiality, that occupies space, that has discontinuous parts, that moves, that can have contact one part with another, that can block motion, that can push one part against another when motion is blocked, that can form groups, that can have emergent pattern of organization, that can form coherent structure, that can develop complexity.

In the case with space a full understanding can be achieved as to what it is and why it exists. Being immaterial it is exceedingly simple. The case is entirely different with matter. It is not known what matter is intrinsically, primally, nor do we have any idea whatsoever about why it exists. Why matter exists is the monumental primary mystery of all reality. Space cannot not-exist because of what it is, but that existence in no way requires the existence of matter. There does not appear to be any reason at all why matter exists. It is just there.

mode-of-being—The manner in which a factor exists. Space exists as the infinite, immaterial, static, three-dimensional extension of place. Matter exists as three-dimensional, extensionally limited, substantial units that can move through space. The existence of a group of material units has three-dimensional extension in space and the substantial three-dimensional extension of the material units, plus it has the organizational factors of the distances and directions between the units. A coherent object has the previous factors of existence plus the manner in which the subunits of the object are held together in structural organization. All the previous situations have a factor of coexistence between the factors of their mode-of-being. The mode-of-being of the continuing-existence of all these situations is ongoing sequentially noncoexistent change—noncoexistent-sequential-difference. Motion and process also have change, noncoexistent-sequential-difference, as the basis of their modes-of-being.

multifactor development—With multifactor development a change in a developmental pathway involves two or more factors. Because the organization of reality is initially complex, virtually all developments involve more than one factor, and often more than one type of development. With codevelopment two or more factors develop together, in concert, often stage by stage. Codevelopment can occur for various reasons, but is often due to a common initiator, or to situations wherein one factor is an aspect of another factor or is consistently a consequence of another factor.

noncoexistent-sequential-difference—Change. Sequential organization in which the components of the sequence are not coexistent. With coexistent-sequential-difference, the sequentially organized components can be, and usually are, existentially independent of one another. With noncoexistent-sequential-difference, the components are sequentially derived, the following parts from the prior parts, and there is thus an existential-dependency relation of what follows on what has gone before. When any particular part of an ongoing noncoexistent sequence is occurring, the following part of the sequence is not there. It has not yet come into being. When that following part is occurring, the prior part is no longer there. It has ceased to exist. The sequential parts of on ongoing change have self-identity because they exist. They are unique and different because they are noncoexistent.

The foundational form of noncoexistent-sequential-difference, of change, is that of spatial continuing-existence (time). Just as the extension of spatial place provides an existential context for the extensional aspects of the mode-of-being of all else that exists, spatial continuing-existence provides an existential-context for all other forms of change. There are two known fundamental origins of change, continuing-existence and motion, with the change that is motion being existentially-dependent on the change that is spatial continuing-existence. All developed forms of change are based on these two.

nonpathway factor development—Nonpathway factor development is the development of a factor from one developmental pathway to another which is in some manner unrelated to the first. In nonpathway factor development, a factor occurs in one pathway in a simpler form and usually at an earlier stage of development of the pathway, and also occurs in a different pathway in a more complex form and usually at a relatively later stage of that pathway. There are two aspects of reality that together result in nonpathway factor development. First, a factor can occur independently in different unrelated existential-pathway-developments. Second, a factor, as it occurs in progressively more complex situations, tends to occur there in progressively more complex, enhanced, form. Factors often show progressive but independent development as they occur in progressively more complex contexts in diverse unrelated existential-pathway-developments, and the roles a factor can play are often tied to its degree of complexity and developmental stage.

With nonpathway factor development there are no roles for existential continuance, neither that of extension nor that of continuing-existence, from one simpler stage across to another more developed stage in a different situation development, nor is there a role for determinate consequent-existence. Thus nonpathway factor development does not play the same type of connected progressive role in the developmental organization of reality that are played by extensional development and change development.

This form of development is usually existentially independent. An exception to this existential independence occurs in cases of requisite existential context relations. For example, the development from the extension of space to the extension of primal matter is a nonpathway factor development in that the reasons substantial extension exists are unrelated to the reasons immaterial spatial extension exists, and yet the extension of primal matter must exist within and conform to the existential context provided by spatial extension. The extension of primal matter is existentially-dependent on the extension of spatial extension, but it is not an existential-pathway-developmental-dependency.

organizational initiation—A form of initiation situation. With an initiation situation factors extrinsic to the initiator play roles in the process of initiation that in part determine the nature of the consequence. With organizational initiation, factors extrinsic to the initiator play organizational roles in the initiation of the consequences. For example, if a moving object had a collision with a stationary object, it would make a difference in the nature of the consequent motion of the stationary object, which direction it would move in, whether the collision occurred more to the right or more to the left on the side of the stationary object oriented towards the moving object. The angle at which the moving object approaches, more to the right or to the left, and the specific place on the stationary object where the collision occurs are aspects of the organization of the situation—organizational aspects that determine, in part, the nature of the consequent motion.

(The initiation of continuing-existence and motion from existence and prior motion is intrinsically unidirectional and thereby there is a factor of organization in the initiation. But because all the organizational factors are intrinsic to the initiator and its consequences, these two cases of initiation are precursors to the more developed situation of organizational initiation.)

place—Location in, part of, or the entirety of existential quantity, as in location in, part of, or the entirety of voluminal spatial extension, and point in, limited sequence of, or all of continuing-existence.

Spatial place is the foundational form. It is pure place—nothing but place. All other cases of place, including the place that occurs with spatial continuing-existence, are developed forms. The place of space itself is the foundation of the development of place involving extension and coexistent-sequential-difference as these factors occur with matter and material organization. The place of spatial continuing-existence is the foundation of the development of place involving noncoexistent-sequential-difference, causal relations, and process. These two developmental sequences interrelate.

The place of spatial extension and the place of spatial continuing-existence provide existential contexts for the forms of place that occur as aspects of the existence of matter and motion. The qualities of the other forms of place conform to the qualities of the spatially based forms.

Developed forms of place occur with patterns of material organization, such as with the place of units within a group, either an open population of separate units or the components of a coherent structure. There is place within coherent matter, and also on its outer surface—material place, which is based on the existential quantity of the matter, place that moves about through immaterial spatial place with the matter. Material place, based on substantiality, occupies immaterial spatial place. As an example think of a check mark on a sheet of paper. The mark occurs at a particular part of the paper, at a particular place on it. If the paper is held up and moved back and forth through space, the mark remains at its place on the material surface of the paper, while the material of the paper, the place on it, and the mark all pass through and momentarily occupy a sequence of spatial place.

Developments-of-origin occur at specific places in the noncoexistent-sequential-difference of the existential-pathway-developments in which those developments-of-origin occur.

population—In general, a population is a group of units. Foundationally, it is a group of primal units of matter. The term usually refers to units which are not coherently bound, such as molecules of a gas or liquid, members of a herd or society, planets orbiting a star, or stars in a galaxy. It can, however, refer to coherent situations, such as the population of atoms of one type distributed within a crystal composed of a different type of atom.

positional orientation—(1) The orientation of the organizational factors of a unit of matter relative to the organizational factors of space, for example, the part of space towards which a specific side of a unit is facing. This occurs also with populations of units and with coherent structure. (2) The orientation of the extensional organization of a motion or a process to the organizational factors of space. (3) The orientation of the organizational factors of one part of matter, a motion, or a process to the organizational factors of another part of matter, another motion, or another process.

primal—That which is primal is the foundation of existence, of reality. That which is primal cannot not-exist—it cannot be created, nor can it be destroyed. That which is primal has always existed, and will always continue to exist. Only space is known for sure to be a primal-form-of-existence. The most primitive form of matter is here considered primal (a) because the intrinsic reasons for the existence of the substantial form-of-existence are independent of the intrinsic reasons for the existence of the immaterial form-of-existence of space, and (b) because space and matter together appear to be sufficient as the basis of all that is evident to exist. But because of the complex factors of the intrinsic nature of matter, and because there does not appear to be any reason for matter to exist at all, its status as a primal-form-of-existence is suspect. The suspicion is that there are probably at least two more primitive, possibly primal, forms-of-existence that together constitute matter.

primal existential factor—*See* existential factor

primary coexistence factors—(*see also* secondary coexistence factors) The factors that constitute the basis of coexistence—the components of the coexistence situation and the simultaneity of their existence. For example, space and units of matter, or two or more units, or a unit of matter and the motion of the unit, or two or more patterns of material organization, either patterns in space or patterns in time. Relations between primary coexistence factors are secondary coexistence factors.

reality referent—That which actually exists to which a term or statement refers. That which exists to which a term or statement is intended to direct the mind. It can also be something that existed in the past or that will exist in the future. Reality referents have been, are, or will be aspects of reality. Unless specifically indicated, the reality referent of a statement is not the concept about something that exists.

role—The role played by a factor is the effect of the factor in a situation due to the intrinsic nature of the factor.

secondary coexistence factors—(*see also* primary coexistence factors) These are the factors of the relative aspect of coexistence. Secondary coexistence factors are factors extrinsic to the primary components of the coexistence. They are the factors of relationship, the factors that occur between the factors which are coexistent.

Coexistence develops, and so do the secondary coexistence factors. The nature of the secondary coexistence factors is usually closely tied to the nature of the primary components and to the nature of the togetherness. The secondary coexistence factors of first stage coexistence, the distance and direction relations between spatial places, are aspects of the extensional aspect of spatial place, and are thus primal aspects of reality. These distance and direction relations exist as the spatial extension between spatial places and as the positional orientation of that in-between extension. Spatial extension is static, existing without any form of change, and there is thus no role for change and newness with spatial places or in the coexistence relations between them. Developed forms of secondary coexistence factors change with changes in the status of the primary components. As an example, the distance and direction relations between two objects of matter exist because the objects exist simultaneously, change when the objects move relative to one another, and cease to exist when the objects cease to exist as objects.

self-identity—What something is intrinsically. What something is in and of itself at any specific point in its ongoing existence. Self-identity includes all the qualities possessed by something that exists, all the factors that play roles in its existence and intrinsic nature, from its primal existential factors, to the factors of the organization of its components, to the features or qualities it has as a whole.

self-organization—When factors intrinsic to a prior stage of development determine organizational factors intrinsic to the following stage, that is self-organization. Self-organization progresses by way of existential-pathway-development. It is self-organization, not disorganization. With the foundational forms, those that occur with the initiation of the organization of the noncoexistent-sequential-difference of continuing-existence and that of motion, it is the creation of orderly sequential relations. With the development-of-origin of emergence, the development of self-organization is not so simple. Emergent pattern of organization occurs with both the coming together of units and with the dispersal of units. The dispersal of a group of units is essentially the destruction of the group quality, essentially the disorganization of the group as a group. This is not what is meant by the term self-organization. The coming together of units creates a form of organization, a group, a primitive form of order. With the origin of emergence, self-organization becomes a form of emergence, the form involving the combining of units.

Self-organization, then, occurs when factors intrinsic to a prior stage of development determine factors intrinsic to the following stage that have some aspect of order. The order can be quite regular, as in the structure of a crystal, or complexly irregular, as in the molecular biology of a living cell. The development from fertilized egg to adult organism, biological evolution, and ecological succession are different forms of highly developed self-organization. Some dispersal events can have factors of self-organization, resulting, for example, in the organization of a nova shock wave and the patterns of organization that sometimes occur in nova remnants.

sequential-difference—*See* coexistent-sequential-difference; noncoexistent-sequential-difference.

sequential enhancement—This is a form of developmental enhancement that occurs with any sequential increase of quantity. In its simpler modes it is the enhancement that occurs in the form of more. The foundational form occurs with increase of spatial extension—greater spatial place has more extension. Spread out from any spatial point location is ever greater extension. At this stage it is a factor of extensional development. Because space is immaterial, spatial extension is continuous, therefore, increasing quantity at its foundational level is continuous and without units, nonnumerical. Sequential enhancement, at its foundation, is continuous.

With spatial extension goes coexistent-sequential-difference—the more extension the more sequential-difference. Sequential enhancement occurs next as a factor of spatial continuing-existence. Here again the situation is continuous, and so is the enhancement. With spatial continuing-existence goes noncoexistent-sequential-difference—the more continuing-existence, the more sequential-difference. There is a stage with the extension of a primal unit of matter, and another with the continuing-existence of matter. These stages are existentially-dependent on the equivalent factors of the spatial context. The next stages occur with continuing motion and the continuing-existence of motion, and with both there is noncoexistent-sequential-difference—the more motion and the more motion continuing-existence, the more of the these two cases of sequential-difference. The stage with material existential quantity will have a factor of continuity involved, and the stages based on material continuing-existence, motion, and motion continuing-existence are all continuous without units, nonnumerical.

All these stages involve foundational forms of sequential-difference. With the extensional cases the parts of the increasing quantity are coexistent, and with the other cases they are noncoexistent. The extensional cases are extensional development and do not involve change, while continuing-existence and motion are cases of change development.

The development of sequential enhancement has next a series of stages involving changes in extensional relations between a moving unit and spatial place and between a moving unit and a static unit, in which the role of sequential enhancement in extensional development and its role in change development occur in one on one relation with each other. The enhancement that occurs at the development-of-origin of emergence, the occurrence of new pattern of material organization due to the motion of a unit and the resulting changes of extensional relations, is a case of sequential enhancement because all the components of the situation are already there together. It is emergence by way of new pattern of material organization by way of sequential rearrangement

In change development there are sequential enhancements where the previous parts no longer exist, such as continuing-existence and motion. There are also cases of sequential enhancement with change development where the previous parts still exist, such as the growth of a crystal or the increase of an animal population. This form of sequential enhancement is accumulative enhancement. Animals have a limited life span, and with the increase of the population there is a turnover of individuals, but when the reproductive rate is high enough there is still an accumulation of individuals and the population increases.

In these cases there occurs a coexistence relation of the newly added component with all the previous components of the sequence. Where there is coexistence, there is relation, and thus combinatorial enhancement. There are cases of change development where sequential enhancement, in the form of accumulative enhancement, plays a role in combinatorial enhancement.

simple change—Change based solely on continuing-existence. Space, matter, and motion, the existential initiators, initiate intrinsic continuing-existence. Simple change then has an existential-dependency development in that motion is dependent on the existence of matter, and matter is dependent on the existence of space. Thus, the simple change that is the continuing-existence of motion is existentially-dependent on the simple change of the continuing-existence of matter, which in turn is existentially-dependent on the simple change that is the continuing-existence of space.

simultaneity—Simultaneity is a situation in which two or more factors exist during the same part of the continuing-existence of space. Coexistence is a situation in which two or more factors exist in space, and do so during the same part of the continuing-existence of space. Coexistent factors continue to exist simultaneously. With coexistence the relation involves the extension of space, and with simultaneity the relation involves the continuing-existence of space. With coexistence there are extensional relations such as direction and distance, while with simultaneity there are relations of the dual occurrence of noncoexistent-sequential-difference such as concurrent occurrence of new part.

The continuing-existence of a factor provides the factor of noncoexistent-sequential-difference which provides the existential context for change in the self-identity of the factor or for change in relation to the factor. The simultaneity of the two cases of continuing-existence of two coexistent factors provides the context for change in the relations between the coexistent factors. Because the development-of-origin for emergence is the change of extensional relations between units of matter, the simultaneity of the continuing-existence of the units plays a required role in providing the existential context for that change. Of course the simultaneity of the continuing-existence of the units, and its role in providing for change between the units, occurs simultaneous with and conforms to the foundational existential context for change provided by spatial continuing-existence.

situation—A situation is any grouping of interrelated factors. A situation can be as simple as the existence of space, or as complex as the entire limitless universe. Because a situation exists, and because it can play a role that influences the nature of a larger situation, a situation is a factor, a developed form of factor. (The universe is a situation, but it cannot be a factor influencing a larger situation because there is no situation larger than the infinite universe.)

stage of development—The nature of a factor between episodes of change. The nature of a factor at different, usually increasing, degrees of its intrinsic complexity. (Often occurs with nonpathway factor development.) The pattern of organization of a situation between organizational changes.

supra-organizational factor—An emergent factor the existence of which is based on at least one factor that is different in nature from organizational factors such as positional orientation or adjacent, direction, and distance relations. In the foundational development of reality, substantiality plays roles in emergence situations that result in the occurrence of supra-organizational factors such as contact, blocking, and push (causal interaction). Coherence (a developed form of interaction) is a more developed form of supra-organizational factor.

Glossary

time—The continuing-existence of space.

Space exists, and it continues to exist. The continuing-existence of space is a consequence of the existence of space. Continuing-existence is a form of change, a form of noncoexistent-sequential-difference, with only the current part of the ongoing change being there, occurring now, the prior part ceasing to exist as the current part occurs and the following part having not yet come into existence. Because continuing-existence is nothing more than continuance-of-being, it is uniform and unidirectional in its occurrence.

Space is the foundation of reality. It provides a place-to-be, an existential context, for all else that exists—or that could possibly exist. Because of what it is, space is infinite, and intrinsically, cannot be otherwise. Spatial place is all the place there is. To not exist in space is to not exist at all. The extensional aspects of the being of all that exists occur within, occupy, and conform to the extensional aspects of space. All that exists other than space has an existential-dependency relation with space and cannot exist independently of space.

In like manner, spatial continuing-existence provides an existential-context for the continuing-existence of all else that exists. All qualities of the continuing-existence of that which exists other than space occur with and conform to the qualities of spatial continuing-existence. The continuing-existence of anything that exists has an existential-dependency relation with spatial continuing-existence, and cannot occur independently of spatial continuing-existence.

Because space is infinite with all parts coexistent, the continuing-existence of space is simultaneous throughout that infinite continuum of place.

transformation point—A point in the existential-pathway-development of a situation at which change occurs that is due solely to the ongoing changing relations between the intrinsic qualities of the components of the situation. At transformation points factors preexisting in the situation can play new roles, but no extrinsic or added component is required to play a role in the initiation of the change. Its occurrence with a single unit of matter moving in relation to spatial place is the origin of the development that progresses to threshold situations.

transformation point initiation—This is an initiation situation, foundationally a type of organizational initiation, in which motion occurs in association with a transformation point and change of some kind is initiated, usually abruptly.

universe—Realistically, the totality of all that exists. It has two known primal or foundational components, space and matter. Space is immaterial, voluminal (three-dimensional) place. The primal intrinsic nature of matter is unknown, but something is-there, exists, that is not immaterial. The universe is infinite because space is infinite.

voluminality—To be voluminal is to have extension in width, height, and depth, to have three-dimensional area as an intrinsic quality of mode-of-being or manner of existence.

Voluminality is a required existential factor. To not have extension in width, or height, or depth, is to not be there to have extension in the other two. To not be voluminal, or have a voluminal aspect of intrinsic existence, is to not exist. To exist at all, there must be some existential quantity of that which exists. Existential quantity is always voluminal or has a voluminal aspect to its mode-of-being.

References

Alexander, R.M. (1994). *Bones: The Unity of Form and Function*, ISBN 0025836757.

Holland, J.H. (1998). *Emergence: From Chaos to Order*, ISBN 0201149435.

Mayr, E. (1988). *Toward a New Philosophy of Biology: Observations of an Evolutionist*, ISBN 0674896653-I (paper).

www.ingramcontent.com/pod-product-compliance
Lightning Source LLC
Chambersburg PA
CBHW060943230426
43665CB00015B/2048